KEYGUIDE TO INFORMATION SOURCES IN

Remote Sensing

KEYGUIDE TO INFORMATION SOURCES IN

Remote Sensing

Edward Hyatt

MANSELL

First published 1988 by
Mansell Publishing Limited, *A Cassell Imprint*
Artillery House, Artillery Row, London SW1P 1RT, England
125 East 23rd Street, Suite 300, New York 10010, U.S.A.
Reprinted 1989

British Library Cataloguing in Publication Data
Hyatt, Edward
Keyguide to information sources in remote sensing.
1. Remote sensing. Information sources
I. Title
621.36"78"07

ISBN 0-7201-1854-9

This book has been printed and bound in Great Britain by
Antony Rowe Ltd, Chippenham, Wiltshire

Contents

Preface

This book was conceived as a reference aid and guide not only for information scientists and librarians, but for researchers and students. An attempt has been made throughout the text to stimulate the individual's interest in remote sensing and its literature, and to equip readers with the necessary means and knowledge to update the information sources themselves, using the methods and publications that are described. To subdivide and separate the information sources in such a varied yet broadly inter-related subject has been problematic, and the divisions that have been adopted should be treated merely as convenient boundaries.

The book identifies and guides the reader towards a considerable number of textbooks and monographs, as well as journals, reports, research materials and trade literature. Advice is given on how to remain up-to-date in remote sensing by utilization of the latest computerized databases and databanks, abstracting and indexing publications, bibliographies, guides to the literature, newsletters and bulletins. The essentially visual nature of the science is catered for by a comprehensive section on world-wide sources of both satellite and aerial photographic remotely sensed data, including relevant published catalogues and manuals of imagery. Audio-visual materials such as slide sets, videocassettes and filmstrips, image mosaics, maps and atlases also receive an in-depth coverage. The book is divided into three parts: Part I is a substantial survey of remote sensing and its literature; Part II is a fully annotated bibliography of the sources mentioned in Part I, and Part III is a selected directory of organizations from around the world.

Throughout the book, numerals enclosed in square brackets ([ ]) refer to cross-references to the bibliographic listing in Part II and the organizational directory of Part III.

I am very grateful to John Cox, currently Engineering Information Specialist in Aston University's Library and Information Service, for his enthusiasm and help with this work. Without his aid this book would have been very much harder to write and his professionalism and friendship have been appreciated.

A great deal of the early history of remote sensing in the introductory chapter was written by Michael Groves of the Remote Sensing Unit in Aston University. I would like to thank him for his useful comments and constant support, together with Lynn O'Connor, who provided help with the structuring of every section of this book (the index in particular). I am grateful to Stephen Birch and Martin Doyle for advice regarding the processing and organization of the main text and index, making a largely thankless task into what was (almost!) a pleasure. Most of my former colleagues in the Unit have been helpful at some time during the writing, and I owe a particular debt of gratitude to Gordon Collins, the Head of Remote Sensing at Aston, for his goodwill and encouragement. I am also grateful to the support staff of the Department of Civil Engineering for taking the time (and a great deal of trouble!) to decipher my handwriting and type the manuscript.

Some of the most important people involved in the compilation of this book I have yet to meet. They are from the organizations to which I sent questionnaires or enquiring letters. I am grateful for the trouble they have taken, and hope they will continue to send me their publications and information so that I may remain up-to-date on their activities.

Finally, I am indebted to the staff of Mansell Publishing Limited, whose patience, comments and advice have proved invaluable at all stages in this books compilation.

Abbreviations and Acronyms

This glossary identifies all of the common, and most of the less common, acronyms and initialisms used in this book. Many of these entries have been excluded from the index for the sake of brevity, although those that occur frequently in the text, or in the opinion of the author are important, may be found in the index.

AAGS	American Association for Geodetic Surveying
AARS	Asian Association on Remote Sensing
ACA	American Cartographic Association
ACSM	American Congress on Surveying and Mapping
AESIS	Australian Earth Science Information System
AIAA	American Institute of Aeronautics and Astronautics
AIT	Asian Institute of Technology
ALS	Australian Landsat Station
AMF	American Mineral Foundation
AMRSA	Atmospheric and Meteorological Ocean Remote Sensing Assembly
APT	Automatic Picture Transmission
ARRSTC	Asian Regional Remote Sensing Training Centre
ARSC	Australasian Remote Sensing Conference
AS	Academia Sinica (Chinese)
ASCE	American Society of Civil Engineers
ASP	American Society of Photogrammetry
ASPRS	American Society of Photogrammetry and Remote Sensing
ASSA	Austrian Space and Solar Agency
AVHRR	Advanced Very High Resolution Radiometer
BARSC	British Association of Remote Sensing Companies
BCRS	Netherlands Remote Sensing Board
BDPA	Bureau pour la Développement de la Production Agricole
BLDSC	British Library Document Supply Centre
BSDSD	British Library Document Supply Division
BNSC	British National Space Centre
BRGM	Bureau de Recherches Géologiques et Minières
CACRS	Canadian Advisory Committee on Remote Sensing
CASLE	Commonwealth Association of Surveying and Land Economy
CCA	Computer and Control Abstracts
CCD	Charged Couple Device
CCEI	Construction and Civil Engineering Index

CCRS	Canada Centre for Remote Sensing
CCT	Computer Compatible Tape
CIAF	Centro Interamericano de Fotointerpretación
CIASER	Centro de Investigación y Aplicación de Sensores Remotos
CLIRSEN	Centro de Levantamientos Integrados de Recursos Naturales por Sensores Remotos
CNES	Centre National d'Etudes Spatiales
CNIE	Comisión Nacional de Investigaciones Espaciales
CNR	Consiglio Nazionale delle Richerche
CNRS	Centre National de la Recherche Scientifique
CONDEPE	Instituto de Desenvolvimento de Pernambuco
COPUOS	Committee on the Peaceful Uses of Outer Space
COSPAR	Committee on Space Research
CRAPE	Central Register of Aerial Photography for England
CRIB	Current Research in Britain
CRS	Committee of Remote Sensing (Vietnam)
CRTO	Centre Régional de Télédétection de Ouagadougou
CSIRO	Commonwealth Scientific and Industrial Research Organization
CSRE	Centre of Studies in Resources Engineering
CZCS	Coastal Zone Color Scanner
CPDI	Fundación Instituto de Ingeniería
CRSTI	Canadian Remote Sensing Training Institute
DAI	Dissertation Abstracts International
DCP	Data Collection Platform
DFVLR	Deutsche Forschungs- und Versuchsanstalt für Luft- und Raumfahrt
DINAGECA	Direcçâo Nacional de Geografia e Cadastro
DIT	Division of Information Technology
DMA	Defense Mapping Agency
DOD	Department of Defense
DOS	Department of Survey
DSIR	Department of Scientific and Industrial Research
DST	Direct Sounding Transmission
EARSeL	European Association of Remote Sensing Laboratories
EARTHSAT	Earth Satellite Corporation
ECA	Economic Commission for Africa
EEA	Electrical and Electronic Abstracts
EEC	European Economic Community
EEI	Electronic Engineering Index
ELV	Expendable Launch Vehicle
EMSS	Emulated Multispectral Scanner
EOPP	Earth Observation Preparatory Programme
EOSAT	Earth Observation Satellite Company
EPO	Earthnet Programme Office

ERA	Environmental Resources Analyses
EREP	Earth Resources Experimental Package
ERIM	Environmental Research Institute of Michigan
ERISAT	Earth Science and Related Information Database
EROS	Earth Resources Observation System Data Center
ERS	Earth Resources Satellite
ERTS	Earth Resources Technology Satellite
ESA	European Space Agency
ESOC	European Space Operations Centre
ESRIN	European Space Research Institute
ESTEC	European Space Technology Centre
ETM	Enhanced Thematic Mapper
FAO	Food and Agriculture Organization
GDDS	GeoRef Document Delivery Service
GDTA	Groupement pour Développement de la Télédétection Aérospatiale
GEMS	Global Environment Monitoring System
GMS	Geostationary Meteorological Satellite
GOES	Geostationary Operational Environmental Satellite
GOMS	Geostationary Operational Meteorological Satellite
GRID	Global Resource Information Database
GSFC	Goddard Space Flight Centre
GTS	General Technology Systems
HCMM	Heat Capacity Mapping Mission
HCMR	Heat Capacity Mapping Radiometer
HDDT	High Density Digital Tape
HRPI	High Resolution Pointable Imager
HRPT	High Resolution Picture Transmission
HRV	High Resolution Visible Scanner
IAPR	International Association for Pattern Recognition
ICA	International Cartographic Association
ICSU	International Council of Scientific Unions
ICTAF	Interdisciplinary Center for Technology Advanced Forecasting
IDB	Interamerican Development Bank
IEE	Institution of Electrical Engineers
IEEE	Institution of Electrical and Electronic Engineers
IFAG	Institute für Angewandte Geodäsie
IFP	Institute Français du Pétrole
IGARSS	International Geoscience and Remote Sensing Society
IGC	Instituto Geográfico e Cadastral
IGM	Instituto Geográfico Militar
IGN	Instituto Geográfico Nacional/Institut Géographique National
IIRS	Indian Institute of Remote Sensing
INEGI	Instituto Nacional de Estadistica Geográfica e Informatica
INPE	Instituto de Pesquisas Espaciais

INTERCOSMOS	International Co-operation in Research and Uses of Outer Space Council
IRS	Indian Remote Sensing Satellite
ISI	Institute for Scientific Information
ISO	Infrared Space Observatory
ISP	International Society of Photogrammetry
ISPRS	International Society of Photogrammetry and Remote Sensing
ISRO	Indian Space Research Organization
ITC	International Institute for Aerospace Survey and Earth Sciences
JPL	Jet Propulsion Laboratory
JRC	Joint Research Centre
JSPRS	Japan Society of Photogrammetry and Remote Sensing
LAPAN	Indonesian National Institute of Aeronautics and Space
LARS	Laboratory for Applications of Remote Sensing
LFC	Large Format Camera
LOC	Library of Congress
LRSA	Land Remote Sensing Assembly
MAFF	Ministry of Agriculture, Fisheries and Food
MIIGAiK	Moscow Institute of Engineers for Geodesy, Aerial Surveying and Cartography
MLA	Multispectral Linear Array
MOMS	Modular Opto-electronic Multispectral Scanner
MOS	Marine Observation Satellite
MSS	Multispectral Scanner
MTFF	Man Tended Free Flyer
NAPL	National Air Photo Library
NASA	National Aeronautics and Space Administration
NASDA	National Space Development Agency
NASM	National Air and Space Museum
NATMAP	Division of National Mapping
NCIC	National Cartographic Information Center
NESDIS	National Environmental Satellite Data Service
NHAP	National High Altitude Program
NLR	National Lucht- en Ruimtevaartlaboratorium
NPOC	National Point of Contact
NRMC	National Resources Management Center
NRSA	National Remote Sensing Agency
NRSC	National Remote Sensing Center
NRSP	National Remote Sensing Programme
NSPS	National Society of Professional Surveyors
NSSDC	National Space Science Data Center
NTIS	National Technical Information Service
OAS	Organization of American States
OEEPE	Organization Européenne d'Etudes Photogrammétriques Expérimentales

ONERN	Oficina Nacional de Evaluación de Recursos Nationales
ORSA	Ocean Remote Sensing Assembly
OS	Ordnance Survey
OSNI	Ordnance Survey of Northern Ireland
OTV	Orbital Transfer Vehicle
PA	Physics Abstracts
PDUS	Primary Data Users Station
PEL	Physics and Engineering Laboratory
PPI	Plan Position Indicator
PSS	Packet Switching System
RBV	Return Beam Vidicon
RCHME	Royal Commission on the Historical Monuments of England
RECTAS	Regional Centre for Training in Aerial Surveys
RESACENT	Remote Sensing Applications Centre
RESORS	Remote Sensing Online Retrieval System
RESTEC	Remote Sensing Technology Center of Japan
ROS	Radarsat Optical Scanner
RRSF	Regional Remote Sensing Facility
RSAA	Remote Sensing Association of Australia
RSC	Remote Sensing Center
RSG	Remote Sensing Group
RSS	Remote Sensing Society
RSU	Remote Sensing Unit
SAC	Space Applications Unit
SAF	Servicio Aerofotogramétrico de la Fuerza Aerea
SAR	Synthetic Aperture Radar
SBPT	Société Belge de Photogrammétrie, de Télédétection et de Cartographie
SCS	Soil Conservation Service
SDD	Scottish Development Department
SDUS	Satellite Data Users Station
SELPER	Society of Latin American Specialists in Remote Sensing
SHE	Subject Headings for Engineering
SIGLE	System for Information on Grey Literature in Europe
SIR	Shuttle Imaging Radar
SLAR	Side Looking Airborne Radar
SLR	Side Looking Radar
SPARRSO	Bangladesh Space Research and Remote Sensing Organization
SPIE	Society for Photo-optical and Instrumentation Engineers
SPO	Science Policy Office
SPOT	Système Probatoire d'Observation de la Terre
SPSE	Society of Photographic Scientists and Engineers
SRIS	Scientific Research Institute of Surveying and Mapping
SRSC	Satellite Remote Sensing Center
SSC	Swedish Space Corporation

SUPARCO	Space and Upper Atmosphere Research Commission
TAC	Technology Applications Center
TM	Thematic Mapper
TRF	Technical Reference File
TRSC	Thailand Remote Sensing Center
UDC	Universal Decimal Classification
UMI	University Microfilms International
UN	United Nations
UNEP	United Nations Environment Programme
UNESP	Instituto de Geociências e Ciências
UNSW	University of New South Wales
USAF	United States Air Force
USGS	United States Geological Survey
VHRR	Very High Resolution Radiometer
VINITI	Vsesoyuznyi Institut Nauchno-Tekhnicheskoi Informatsii
WAPI	World Aerial Photographic Index
WEFAX	Weather Facsimile
WISI	World Index of Space Imagery
WMO	World Meteorological Organization
WRS	World Reference System
WTI	World Transindex
WWW	World Weather Watch

PART I

Survey of Remote Sensing and Its Literature

1 Introduction

1.1 The Early History of Remote Sensing

The term 'remote sensing', as applied in the present-day context, first came into existence during the development of spaceborne remote sensing systems in the early 1960's. The systems were exemplified by the excellent photographic results obtained from the Mercury and Gemini missions, which in themselves paved the way for the non-photographic sensors of the 1970's. The rapid changes that led to the birth of this new terminology were characterized by Colwell (1979), who divided the development of remote sensing into two areas of interest: pre-1960 and post-1960. Prior to 1960, aerial photography was the only operational imaging system, and itself developed in tandem with the advances made in photographic film, camera and platform technology.

This parallel development can therefore be strictly traced back 2,300 years to Aristotle's experiment with a 'camera obscura'. However, within the bounds of photographic images from the air, it is recorded that the first aerial photograph was taken in 1859 by Gaspard Felix Tournachon. This unique image was taken from a balloon over the village of Petit Bicetre near Paris (ostensibly for mapping purposes). Tournachon, who later known became known as Nadar, had no choice but to develop the film whilst still airborne, although by 1871 Maddox had developed a gelatin emulsion of silver halide grains so that an image could be developed at a later time.

During the late 19th and early 20th century much of the original work was undertaken which provided the basis for the development of non-photographic sensors and colour films, as well as the platforms themselves. Notable examples of this type include De Hauron's work on three colour separations using red, yellow and blue pigments (1895) and Heinrich Hertz's discovery in 1889 that solid objects reflect radio waves. Of direct relevance to remote sensing at the time however, were the first aerial images to be taken from an aircraft. These motion picture images were obtained by Wilbur Wright on 24 April 1909 whilst flying over Centonelli in Italy. Prior to the refinement of the aeroplane, a variety of platforms had been used, ranging from balloons and kites to a 70 gm camera strapped to a pigeon's chest. However, once established, the combination of aeroplane and camera developed rapidly, especially during the First World War. It did not take long before both sides in this conflict discovered the advantages of photo-reconnaissance; thus in 1915 the first practical aerial camera was introduced by J. T. C. Moore-Brabazon of the RAF. Improvements in this form of military intelligence were prompted by increasing demand for its service, a clear example of the enthusiasm shown for aerial photography being the acquisition and development of prints covering the entire battlefront during the Somme offensive in 1916 (Brookes, 1975).

The value of aerial photography was amply demonstrated during the war, although few training procedures had been established and few aircraft and aerial cameras were available for training purposes. During the inter-war years, though, development did continue, especially in the science of photogrammetry. The pioneering work had been completed by Aime Laussedat in the mid 19th century, but it was the advent of stereo plotters, such as the Zeiss multiplex, utilizing vertical aerial photographs, that moved this field into its present realms. The mapping aspect of aerial photography characterized by photogrammetry was pursued during this period by survey and planning bodies, especially in the U.S.A., where the U.S. Geological Survey (USGS) and Tennessee Valley Authority became chief exponents of aerial survey techniques. With increased usage of photography came the development of guidelines and strategies for interpretation of the imagery, one of the first books on air-photo-interpretation being published in 1922.

Other major developments during the inter-war years included experiments in the early 1930's with an infrared-sensitive film (PAN-K) which retained good resolution from 20,000 ft over Rumford, Maine. On the basis of Mannes and Godowsky's work in 1924 on multiple-layer colour film, Kodak marketed the first colour film (Kodachrome) during 1935 in a 16mm format. This prompted General George Goddard of the U.S. Army Air Corps to test the new film in an aerial photographic capacity; however, the processing and haze problems associated with this emulsion were not solved until Kodak introduced Kodacolor Aero Reversal Film in 1942. This advance in colour film emulsion was accompanied by developments in multispectral black-and-white photography exemplified by a twin-lens hand-held aerial camera and viewer introduced by a German company in the early 1930's. The applicability of multispectral photography was vastly increased by new films and colour additive viewing systems which corrected for mis-registration of the composite images. Such camera and viewing systems were given their first major trials on the Apollo and Skylab space missions, proving their worth for a variety of applications.

The next strides to be taken in remote sensing technology were a direct result of the Second World War, the main advance being observation in wavelengths beyond the visible regions of the spectrum. At the beginning of the conflict, both Allied and Axis powers had well-established photo-reconnaissance procedures. Indeed British intelligence scored two notable successes in detecting the build-up for an intended invasion of the British Isles and in identifying the launch site for V-1 rockets later in the war. However, it was the development of colour-infrared film and radar that provided the main impetus to remote sensing as it is considered today. The effect of the advances made in Germany with rocket and turbojet technology cannot be ignored or overstated, but of more immediate interest to the remote sensing community as such was the refinement of colour-infrared film.

In 1941 the United States government set up a National Defense Research Committee to investigate the spectral properties of camouflage. It had previously been recognized that the peak vegetation reflection in the infrared region beyond 0.7 μm and the absorption band in the blue part of the spectrum (0.67-0.68 μm) could not be emulated in camouflage paint. Thus it only became a matter of time before an infrared emulsion was incorporated into a colour film. This particular honour goes to Spencer and Marriage working at the Kodak Research Laboratories in Great Britain, who in 1941 produced the first infrared film with colour processing. The next step involved modification of Kodacolor film to respond to the green, red and infrared regions of the electromagnetic spectrum, the result being the first 'true false-colour' film, which eventually became available to the military in January 1943, as Kodak Ektachrome Infrared Aero Film: Type 8443 (Aero Kodacolor Reversal Film). The modern equivalent of this film (Type 2443) has a special niche in remote sensing applications for the assessment of vegetation stress owing to its enhanced contrast and much debated ability to detect pre-visual change in infrared reflectance before a visual change occurs. Colwell pioneered this application of colour-infrared film during the late 1950's; however, there has now been some movement towards the use of colour-infrared negative film as an applications tool. For a concise overview of both the development and properties of this type of film the reader should refer to *The surveillant science: remote sensing of the environment* [187] and the *Manual of remote sensing* [185].

Based on developments at the turn of the century with radio detection and ranging (RADAR), both Britain and the United States had viable systems for ship and plane detection during the war, the first successful airborne imaging radar, better known as the Plan Position Indicator (PPI), having been developed in Britain during the inter-war period. A refinement of this so-called 'real aperture radar' was the synthetic aperture radar (SAR), in which the antenna is synthetically created permitting fine resolution with long wavelengths and low power. Accordingly, post-war development in radar concentrated on SAR and Side-Looking Airborne Radar (SLAR), a real or synthetic aperture system providing a scan capability parallel to the line of flight. The first complete SLAR system, developed in 1954, was designed for use in high-altitude aircraft used by the U.S. Strategic Air Command. Meanwhile, the U.S. Air Force funded development of a system for the generation of high-resolution imagery. However, this and subsequent SLAR systems were not extensively used for terrain survey until NASA sponsored a programme for this purpose in 1964. The ability to penetrate cloud and surficial sediment has made SLAR systems popular for geological mapping of large-scale structures as well as for drainage mapping and sea ice assessment. A thorough overview of radar applications can be found in Lillesand and Kiefer (1979) and Sabins (1978).

It is worth mentioning the principle behind synthetic aperture radar, since it is now the most widely used system for remote sensing purposes. The mechanism involves the resolution of the difference between the transmitted signal and its corresponding scattered return signal (Doppler shift). The result is a high spatial resolution required for terrain interpretation, but it was not until the late 1950's that a suitable method for analysing the scattered signal was refined, and it is only in the past fifteen years that its advantages have gained widespread acceptance. A good example of the confidence shown in radar was the Shuttle Imaging Radar (SIR), a synthetic aperture system carried on two Shuttle missions in 1981 and 1984 (SIR-A and SIR-B).

Radar was just one of the non-photographic sensors refined during the post-war period. Another notable addition to the systems and sensors of remote sensing was the development of 'thermal' infrared detectors. This technology paved the way for the 'scanner systems' that have characterized the short period of space remote sensing. In 1962 the military had already spotted the potential of an infrared sensor for civilian remote sensing. This potential could not have been fulfilled, however, without the original work undertaken by Czerny, who devised the first successful infrared imaging device in 1920, and Krinov, who experimented with spectral reflectance properties. Krinov's work led to the formulation of radiation models to aid in the interpretation of infrared images and also laid a considerable amount of the groundwork for the development of radiometers and spectrometers able to sense specified wavelengths over the whole spectral range. As a result, thermal linescanners recording digitally are used for a variety of purposes including crop type assessment and detection of water stress (Curran, 1985).

In addition to thermal-infrared sensing, other developments in the post-war period included ultraviolet films and laser devices. Astronomers were the first to realize the advantages of UV-sensitive film at the turn of the century; but it was not until the late 1960's that the technology was applied earthward. It was noted that carbonate rocks, concrete and metal retained high reflectance values in the UV region (3-3.5 μm), thus allowing their differentiation from other surface features. Another intriguing usage for UV film has been the detection of harp seals in Newfoundland; because their fur absorbs UV radiation they appear dark against the snow in photographic images (Holz, 1985). Wider applications have been demonstrated for laser and luminescence technology, which is the youngest technique to have been applied in a remote sensing capacity. The important advance in this field was made with the realization that use of short-wavelength lasers could broaden the spectrum of application because of the phenomenon of laser-induced fluorescence. Subsequently, a laser-fluorosensor was introduced which is accurate enough to define the extent of an oilslick for use as evidence in a court of law. The other major tools of remote sensing laser technology are light detection and ranging (LIDAR) systems. Based on the principle that the time between a transmitted and a scattered return laser pulse can be related to the amount and type of molecules encountered, both systems have been used to monitor water and air pollution (Measures, 1984).

The ability to monitor the Earth's surface in a variety of wavelengths, although invaluable, could only be applied to relatively small areas of the Earth at one time. Before the first manned orbital space missions, the maximum altitude that a manned aircraft could achieve was approximately 70,000 ft., this honour belonging to the Lockheed U-2 military reconnaissance aircraft of the type piloted by Francis Gary Powers and shot down over the U.S.S.R. in 1960. The possibilities of using rocket-borne cameras for earth observation had been recognized over fifty years earlier when Alfred Maul patented a gyroscopically stabilized camera for mounting in a rocket that was rather similar in appearance to a large firework! Although Maul's rocket attained a maximum altitude of only 2,600 ft., the idea had at least been tested. Thus it was that the possibilities offered by the National Aeronautics and Space Administration (NASA) space programme for remote sensing were eventually acknowledged. The first orbital photographs (167 frames) were taken from an unmanned Mercury-Redstone 2 orbiter on 31 January 1961. The photographs of North Africa amply demonstrated the synoptic capabilities of this type of photography, the images even being used to produce a series of geological maps of the Sahara desert. During the subsequent manned Mercury missions further high-quality 35mm and 70mm colour photographs were obtained by each astronaut. The response to these images was highly encouraging, so two experiments were designed for the second-generation Gemini programme.

Experiments S005 (Synoptic Terrain) and S006 (Synoptic Weather) produced more than 1,100 high-quality colour photographs during a total of ten missions. As part of the Gemini IV mission (3-7 June 1965), thirty-nine exposures with six-second intervals provided the first stereoscopic coverage from space, which extended from California to Texas. Some months later, the crew of Gemini V exposed the first infrared film from orbit; however, the Ektachrome Aero Infrared Film that they used proved too thick for the magazine, and as a result the emulsion was badly scratched. The two crew members of Gemini IX (3-6 June 1966), photographing Peru simultaneously, managed to attain coverage for almost three-quarters of the country in three minutes. These particular results were appreciated, since only one-quarter of Peru had been photographed from the air during the previous fifty years. In September of that year the Gemini XI orbiter achieved a record altitude of 739 nautical miles, acquiring photography at different scales. The application of these early space images had already been spotted by the geological community and it was the USGS that expounded a general plan for the development of what eventually became the Landsat series of satellites.

This increasing interest shown in orbital imagery by terrestrial and marine scientists prompted the next step in space remote sensing. A combination of multispectral remote sensing systems and high-altitude space vehicles was an obvious progression. Therefore, the S065 Multispectral Terrain Photography

Experiment was carried on Apollo 9 in March 1969. The system consisted of four Hasselblad cameras mounted together, each having a different film/filter combination allowing coverage in the yellow, green, red and infrared (the spectral bands planned for the ERTS-A/Landsat-1 Multispectral Scanner). The images obtained proved very promising for earth observation, but since the Apollo missions were directed towards a Moon landing, a different platform had to be considered. The obvious choice was NASA's Skylab programme, although some poor-quality 35 mm and 70 mm photography was acquired in 1975 during the joint U.S.-U.S.S.R. Apollo-Soyuz Test Project. The U.S.A.'s first space station was launched on 14 May 1973 into a near-circular orbit (435 km) carrying the Earth Resources Experiment Package (EREP). Two cameras were carried, the S190A multi-spectral camera system and the S190B Earth Terrain Camera. S190A consisted of six Hasselblads which covered the same ground area with six different film/filter combinations over the 0.41-0.9 µm region. S190B was a single camera of focal length 127 mm covering the 0.4-0.88 µm range. However, more important to the future of remote sensing were the infrared spectrometer (S161) and thirteen-channel multispectral scanner (S192) also carried. These systems, although not perfect, signalled the end of conventional orbital photography, until the Shuttle Large Format Camera (LFC) and Metric Camera, but acknowledged an era of synoptic coverage with scanner data which had already 'taken off' on 23 July 1972 with the launch of the first orbital remote sensing satellite, ERTS-1, which was later renamed Landsat-1.

1.2 Review and Preview of Spaceborne Land Remote Sensing Systems

It is not the intention of this section to dwell on the developmental history of satellite remote sensing, and the following account presents a brief précis of those spaceborne remote sensing systems that have been instrumental in charting the path of future earth observation missions. Technical parameters, such as the satellite's launch date and the bandwidths and functions of the onboard sensors, are described in numerous publications, including the *Manual of remote sensing* [185] and *Remote Sensing Yearbook* [6]. Many of the systems mentioned incorporated a range of sensors, but only those of direct interest to remote sensing are outlined. Future systems are presented in less detail, as the technical specifications of these missions are liable to fluctuate widely before a consolidated agreement as to payload type and form has been attained.

Landsats-1, 2, 3, 4 and 5

The first Earth Resources Technology satellite (ERTS-1) was launched by NASA in July 1972 and renamed Landsat-1 upon the launch of Landsat-2 in January 1975. The satellite was designed to collect data about the Earth's surface and carried a payload that consisted of two sensors: the Return Beam Vidicon (RBV) Camera and a four-band Multispectral Scanner (MSS). A fifth band in the thermal infrared region was added in Landsat-3, together with an updated RBV system with twice the resolution of the previous system and a different imaging configuration.

Landsat-4, which was launched in July 1982, heralded the advent of a second-generation Landsat series, as it carried a seven-band Thematic Mapper (TM) with increased spectral sensitivity and spatial resolution in addition to the MSS payload carried on Landsats 1 and 2. Unfortunately the improved TM sensor was partially closed down after only 213 days owing to a power failure. However, the launch of a modified Landsat-5 in March 1984 re-established a fully functioning TM sensor.

A further generation of Landsat satellites has been proposed by EOSAT [741] following its commercial takeover of the Landsat programme. At the time of writing EOSAT proposes to use an Advanced TIROS-N/Defense Meteorological Satellite Programme spacecraft for Landsat-6. The spacecraft, which will probably be launched by a U.S. Titan II Expendable Launch Vehicle (ELV) early in 1989, will include an Enhanced Thematic Mapper (ETM) that duplicates the current TM bands and features an additional 15 m panchromatic channel in the 0.5-0.9 μm wavelength region. It is further proposed that the same bands as the earlier MSS of Landsats 1, 2, 4 and 5 will be provided at 60 m resolution by the Emulated Multispectral Scanner (EMSS). EOSAT had previously favoured employing OMNISTAR, a modular long-life spacecraft which can carry a variety of payloads, together with fuel supplies that would allow it to change to a low orbit for equipment servicing by the Shuttle for Landsat-6. The OMNISTAR concept will most probably be employed for Landsat-7, ensuring data continuity by its twenty-year life, and providing a flexible system that can respond to developments in commercial remote sensing sensors and applications. A further stage is envisaged that will take the Landsat series into the next century and its fourth generation. It is proposed that the Landsat-H satellite will carry a payload that includes Multispectral Linear Array (MLA) 'Smart Sensors', a 5 m High Resolution Pointable Imager (HRPI), L, C and X Band SAR and an active optical sensor for night imaging and atmospheric calibration.

Système Probatoire d'Observation de la Terre-1, 2, 3 and 4 (SPOT)

The first French remote sensing satellite, SPOT-1, was placed in orbit on 22 February 1986 at 01.44.35 a.m. by the Ariane-1 launcher. Spatial resolutions of 10 m and 20 m respectively in one panchromatic and three multispectral modes, allied with a revisit capability that can provide stereoscopic imagery and which will allow an area to be imaged every five days, provide SPOT-1 with unique imaging capabilities.

The current SPOT programme has been defined into the 1990's. SPOT-1 and its successor SPOT-2 have two-year design lives and are essentially the same, although the HRV scanners on SPOT-2 will have Thompson-CSF linear arrays instead of the current Fairchild CCD. Further payload modifications may include the addition of a Doris satellite positioning system accurate to 10 cm, and a Poseidon ocean monitoring package that employs an altimeter microwave sounding unit and laser retroreflector for its measurements. The French government gave the go-ahead for the continuation of the SPOT programme in 1985. SPOT-3 is expected to have a four-year life and should replace SPOT-2 towards the end of 1990. SPOT-4 will be available for launch in 1991 as a back-up to SPOT-3; however, launch will be in 1994 if no problems occur with SPOT-3. The sensors on these latter-generation series will be modifications and improvements of the former payload. A 20 m mid-infrared vegetation channel (1.5-1.7 μm) will be introduced, thus providing four MS bands. Data compression techniques will allow transmission to Satellite Data Receiving Stations at the same rates used for SPOT-1 and 2 transmissions. The second MS band will relay both 10 m panchromatic and 20 m MS data, which will considerably aid geometric registration between the different channels. A wide-angle imaging instrument with a 900 m resolution is planned for global vegetation monitoring. This sensor will utilize the same bandwidths as the HRV's and its image swath will be almost 2000 km, thereby facilitating daily global coverage when used in conjunction with the satellite's onboard recorders.

U.S.S.R. and Military Satellites

The U.S.S.R. operates a range of remote sensing satellites as part of its Cosmos series. Most of these missions use photographic sensors to obtain earth resources data, but in September 1983 the Soviets launched Cosmos 1500, their first satellite incorporating Side Looking Radar (SLR). A year later Cosmos 1602, another radar-equipped resources satellite, was put into orbit. In each case some form of low-resolution multispectral facility is carried to complement the 1.5-2 km resolution radar.

The larger part of the Cosmos series consists of military reconnaissance satellites, and up to 100 of these are launched by the U.S.S.R. each year. As may be expected, very little is known about these missions; however, the U.S.S.R. appears to be launching fourth-generation, high-resolution photographic reconnaissance satellites quite regularly. Early models in this series were very short-lived, remaining in orbit for only one to two weeks before burning up on re-entering the atmosphere and jettisoning their film pods for mid-air or seaborne retrieval. Those in the latter series are more durable and appear to be capable of remaining in orbit for longer periods of up to six months. Beginning in 1985 the Cosmos series was expanded to include electronic reconnaissance platforms in the 1536 and 1544 series that may have orbital lives of up to sixty years.

The United States regularly launches manoeuvrable photographic reconnaissance satellites in the Big Bird (USA-2) and USAF series. These expendable reconnaissance packages have lives of four to six months, although the Big Bird missions occasionally include payloads consisting of electronic imaging satellites that are launched once the parent satellite has attained orbit. Such spacecraft are probably related to the KH 11 'keyhole' series, which carry advanced sensor packages that can achieve high-resolution imaging in combination with onboard digital processing. The Space Shuttle will also fly assignments such as the

Department of Defense (DOD) Spacelab (June 1989), which probably includes reconnaissance packages and related experiments.

Metric Camera and Large Format Camera (LFC)

The European-sponsored Metric Camera and the U.S. Large Format Camera missions were flown on the Space Shuttle to test the potential of high-resolution spaceborne photography using modified high-altitude aerial cameras. The Metric Camera mission was largely successful, obtaining nearly 1,000 images, most of which could be viewed stereoscopically. Some technical problems did occur with black-and-white image acquisition and the lighting conditions were such that this European initiative did not actually acquire very high-quality photographs of Europe! Unfortunately bad weather conditions over North America and Europe resulted in much the same problem during the LFC mission. However, given the large area covered by each LFC picture, the cloud cover is not always excessive and large areas of the ground are sufficiently clear to demonstrate the effectiveness of the system.

Modular Opto-electronic Multispectral Scanner (MOMS)

The MOMS system has been flown in the Space Shuttle three times, once in 1983 and twice in 1984, but the second attempt to fly the scanner on 3 February 1984 was unsuccessful owing to operational problems. The modular arrangement of this system, which was developed at the Deutsche Forschungs- und Versuchsanstalt für Luft- und Raumfahrt (DFVLR) [609], allows it to perform target specific data capture in a variety of bandwidths. A 10 m resolution stereo-imaging capability is currently being developed at the DFVLR and it is also planned to extend the sensor's spectral range with mid-infrared and thermal infrared channels for future missions.

Shuttle Imaging Radar A and B (SIR)

The SIR-A experiment, which was flown in March 1981, collected optical image data for more than eight hours. The mission objectives were to examine areas that other types of sensor could not, such as densley forested or cloudy regions of the Earths surface, and to test the suitability of imaging radar over a wide range of geoscientific applications. The experiment covered an area in excess of 10 million km^2 and ably demonstrated the useful penetrative qualities of radar. SIR-B, a progression from the earlier package, was flown on the same Shuttle mission as the LFC. An identical L-band horizontally polarized wavelength was used, but the imaging sensor was able to acquire both optical and digital material at a range of radar incidence angles. Unfortunately data relay problems meant that only one-fifth of the projected SIR-B coverage was obtained; however, the two experiments were very useful for the determination of design parameters in future radar systems.

Seasat

The SIR missions were a follow-on from Seasat, NASA's first spaceborne radar experiment. The Seasat payload comprised five different sensors, including a 25 m resolution L-band SAR, capable of imaging the Earth's surface despite adverse atmospheric conditions. The mission was launched on 26 June 1978 and relayed a vast amount of high-quality data until its premature failure, owing to a massive electrical short circuit, on 10 October 1978.

Heat Capacity Mapping Mission (HCMM)

The HCMM was launched in 1978 and operated until its retirement on 30 September 1980. Visible/near-infrared and thermal infrared data were respectively acquired at resolutions of 500 m and 600 m by its sensor, the Heat Capacity Mapping Radiometer (HCMR). The circular Sun-synchronous orbit of the satellite was timed to take maximum advantage of the temperature variations in the northern hemisphere. The day and night temperature difference images that were obtained aided in the determination of the scanned surfaces' thermal inertia and provided data on the feasibility of using thermal infrared imagery for environmental monitoring.

Coastal Zone Color Scanner (CZCS)

One of the most long-lived of the NASA experimental resources satellites of the late 1970's was the Nimbus-7 CZCS. This near-polar orbiting oceanographic and meteorological satellite operated from 24 October 1978 until September 1984. The six-channel 800 m resolution CZCS was the main earth resources instrument amongst the nine sensor packages that constituted its payload. Five bands of the CZCS were designed to measure water colour, whilst one was reserved for sea surface temperature data.

International Space Station

The somewhat futuristic concept of a permanent manned space station in low Earth orbit is nearing the end of its design and specification stage at the time of writing. Current trends are towards a single-keel platform with various manned and unmanned modules provided by the United States [772], the European Space Agency (ESA) [495] and the National Space Development Agency (NASDA) [643]. Canada has been contracted to supply a mobile servicing platform that will carry out maintenance, amongst its other functions. It is envisaged that construction of the International Space Station will begin in 1992, with completion and operational status being attained by 1994 after twelve Shuttle flights.

The original concept of the Space Station was for a dual-keel core with six manned modules. This system, which currently seems somewhat unrealistic, would have been established between 1993 and 1995 after seventeen Shuttle flights. The conversion of the single-keel model to dual-keel capability by the end of the 1990's has been postulated by NASA. However, it is unlikely that twelve flights would be enough for the initial core and a figure nearer twenty flights may be more realistic in the two-year period. Furthermore, as the Shuttle has yet to have more than nine operational flights in one year, the time-scale may also be somewhat optimistic.

The core platform of the International Space Station should exist as a single entity. However there are several other associated satellites that fall under its auspices which should be considered as an integral part of the whole Space Station concept. Perhaps the most important of these in terms of earth observation are the two proposed polar platforms. These permanent facilities, which are to be supplied by the U.S. and ESA, will have complementary Sun-synchronous morning and afternoon orbits in order to provide complete daily global coverage. A variety of remote sensing packages such as the Ocean Remote Sensing Assembly (ORSA), Land Ocean Remote Sensing Assembly (LRSA) and the Atmospheric and Meteorological Ocean Remote Sensing Assembly (AMRSA) are to be included as part of a varied payload. In-orbit servicing every two to three years for repair and addition of new sensor packages has been proposed using the Shuttle system. One of the problems inherent in this design is the operating altitude of the Shuttle. The polar platforms would be in a relatively high geosynchronous orbit and the transfer of the satellites between this and the proposed 436 km low Earth orbit of the International Space Station core may be problematic. The same sort of manoeuvrability that is being developed for the Eureca system (which is described later) is implied, although the idea of a remotely controlled Orbital Transfer Vehicle (OTV) that would bring the geosynchronous data relay satellites down to low Earth orbit is also under consideration. If the polar platforms are to develop in synchrony with the rest of the Space Station, they must be launched at around the same time. Initial thinking favoured employing the Shuttle for these launches, but in consideration of the time delays discussed elsewhere, it is now more feasible to use ELV's, such as Ariane 5, or the U.S. Titan rocket, which is also being increasingly regarded as a launcher for some sections of the core station. One big advantage of the modular construction and requirement for sensor package servicing and modification is the launch cost. Previously whole satellites have been launched by Shuttle and ELV; however, with this system it is possible to launch only payloads consisting of various packages, as the platform infrastructure will already be in space.

Various satellites and payloads are being considered as part of the European Columbus programme, which will contribute one module to the core station. The ESA originally wanted their module to be an independent entity operating partly within and partly outside Space Station constraints. They have now accepted an integrated fixed core module, but plans are being considered for the construction of a similar external module, which could be maintained as a Columbus Man Tended Free Flyer (MTFF), a small space vehicle capable of docking with both the Space Station and Shuttle.

The International Space Station concept seems to have come of age and can now be realized, although there are some problems. Most of these, which are political and legal in nature, could prove more damaging to the Space Station concept than any lapse in technological development, but hopefully the project will develop as a co-operative multinational exercise without recourse to what are best described as petty nationalistic ideals.

Eureca

The Eureca retrievable platform is in essence a precursor to the manoeuvrable MTFF mentioned earlier; in fact the MTFF may actually be a modified Eureca platform. The Eureca system is being developed at MBB-ERNO for the ESA and will become operational early in 1988. The carrier will be launched in the Shuttle and deployed by the orbiter at around 300 km; it will then propel itself into an operational orbit at around 500 km. The length of each mission should be about six months, after which the Eureca platform will drop to Shuttle altitude for recovery and return to Earth. If no retrieval facility is at hand for relocation, the satellite can orbit for up to nine months, although its operational phase will still remain at six months. A variety of packages are envisaged, including a high-resolution earth observation payload that will carry a variety of sensors suited to short-lived mapping missions in specific application areas.

Earth Resources Satellite (ERS)

The ERS-1 satellite is another European initiative that is expected to be the precursor of a series of operational European remote sensing satellites. A variety of sensors are anticipated as part of what is essentially an ocean and ice surveillance package. An all-weather C-band, VV polarized SAR will be part of the imaging mission, and visible and infrared scanning radiometers will be used to determine sea-surface temperatures and atmospheric vapour levels. Further satellites in the ERS series, which are planned for the 1990's, will continue the meteorological, climatological and imaging missions of ERS-1, but may incorporate payload additions and sensor modifications that will be tuned towards land remote sensing.

Radarsat

A 25 m resolution SAR and Radarsat Optical Scanner (ROS) are amongst the sensors proposed for the Canadian Radarsat platform. The projected launch date of Radarsat is in late 1990 and a wide variety of applications are envisaged because of its sensor's cloud penetrative ability.

Other Satellites

The Japanese will also launch a satellite that is geared towards oceanographic observation in late 1987. The Marine Observation Satellite (MOS-1), to be operated by NASDA, will have a variety of medium- to low-resolution oceanographic sensors. MOS-2 should be launched in 1989 or 1990, and five more satellites in the series are scheduled for launch before or during 1998, with the aim of maintaining data continuity into the next century. Further projected satellite launches include the Brazilian remote sensing satellite, a Brazilian Space Research Institute [532] initiative, in 1991 and the Indian Remote Sensing (IRS) satellite launch sometime in 1987. Amongst other sensor packages, both of these satellites employ medium- to high-resolution CCD camera arrays.

1.3 Review and Preview of Spaceborne Meteorological Satellites

It is quite difficult to remain up to date on the status of meteorological satellites (Metsats) as their very nature and mode of operation are quite flexible. Determining the current positions of these spacecraft, and especially the geostationary platforms, is problematic as their locations are seasonally adjusted to take account of weather variations and to provide cover for malfunctioning satellites, or until the launch of further satellites in a series. The best way to remain informed of Metsat status is to consult the updates on these platforms in the newsletters listed in Part II, or to contact one's local meteorological office, which will be aware of the current locations of operational systems.

TIROS-N series

NOAA currently aims to maintain two polar orbiting satellites in orbit at all times, thus allowing an overpass of all areas approximately every six hours. Data are either broadcast directly to ground receiving stations, or are recorded on board for transmission at a later date. Visible and near-infrared imaging and atmospheric sounding data are provided by the satellites, which also act as Data Collection Platforms (DCP), relaying broadcasts from remote spacecraft. Three of the current platforms have an imaging capability, although the Advanced Very High Resolution Radiometer (AVHRR) on NOAA-6 is only at 50% capacity owing to power supply problems and at the time of writing the satellite has been switched to standby following its reactivation on 13 June 1984 as a result of the failure of NOAA-8. NOAA-9 and NOAA-10 are operating satisfactorily and although there are problems with the infrared sounder on NOAA-9 the imager continues to provide useful data. The launch of NOAA-H is scheduled for 1987, with three further satellites, NOAA-D, NOAA-I and NOAA-J, proposed up until 1989-1990. The next generation of polar orbiters, to be launched in 1990, 1991 and 1992, may carry an additional payload that includes the Ocean Color Imager.

Meteor-2 series

The Meteor-2 series is maintained by the U.S.S.R. as the basis of its operational polar orbiting meteorological satellite service. The operational series are in a 900 km orbit and provide imagery products very similar to those of the NOAA polar orbiters. Experimental series are in a low 650 km orbit, but the data transmission frequencies of some of these platforms are unknown. One to two satellites are launched each year, and it is expected that the series will be operational up to 1990. Modifications to the next generation of Meteor satellites may include higher orbits, increased resolution and a microwave sensor capability.

Geostationary Operational Environmental Satellite (GOES)

The World Weather Watch (WWW), a global observation system initiated by the World Meteorological Organization (WMO) [503], requires five operational geostationary satellites positioned around the equator at 70-75° intervals. The U.S. contributes two satellites to this pool, GOES East and GOES West, and the U.S.S.R., ESA and Japan all contribute one satellite to this Geostationary Meteorological Satellite (GMS) programme. The system is ideal, but unfortunately only one U.S. satellite is currently operational and the Soviet platform has yet to be launched. India operates Insat-2, a geosynchronous meteorological and communications platform equipped with a Very High Resolution Radiometer (VHRR), independently of this programme. Insat-2 provides half-hourly pictures of the Indian sub-continent at resolutions of 2.75 km (visible) and 11 km (infrared).

The U.S. GOES System currently has only one operational platform, GOES-6, following the destruction of GOES-G (which would have become GOES-7 had it achieved a successful orbit) on launch in May 1986. At any one time NOAA prefers to keep two GOES satellites in orbit and the agency now faces serious problems if failure should occur with GOES-6 before the launch of GOES-H in 1987. GOES-3 and GOES-5, which provide data services for GOES-6, are still capable of broadcasting Weather Facsimile (WEFAX) imagery and GOES-2 supports these services. GOES-4 has been moved to 44° W to act as a data collection and broadcast platform for Meteosat-2. The next generation of geostationary satellites, GOES Next, I, J, K, L and M, will ensure data provision into the late 1990's and are improved versions of the earlier systems.

Geostationary Meteorological Satellite (GMS)

The Japanese GMS-3 is operated by the Japanese Meteorological Agency; it provides imagery at two resolutions and also acts as a DCP under the WWW programme. GMS-4 is due to be launched in 1987 and GMS-5 in 1995. These will be followed in 1998 by GMS-E, the prototype for a series of high-resolution meteorological satellites.

Geostationary Operational Meteorological Satellite (GOMS)

The Soviet GOMS is due to occupy a position at 70° E; its projected launch date is sometime in 1987 and it will carry a payload complementary to the present GMS and GOES series.

Meteosat

Two satellites in the Meteosat series have been launched to date, and a third, Meteosat P-2 (a modified prototype), is due for launch in 1987. Upon launch, the data collection antenna of Meteosat-2 was found to be faulty and the satellite had to rely on Meteosat-7, its predecessor, for DCP capabilities. Data relay facilities for Meteosat-2 broadcasts have now been assumed by GOES-4, which has been moved to a position at 40° W to allow continuation of the DCP mission. The orbit of Meteosat-1 has now decayed to such an extent that it is out of data transmission range.

Many other countries are launching or plan to launch various remote sensing satellites in the future. Fuller descriptions of some of the systems mentioned above are given in a variety of publications including *Principles of remote sensing* [186], *Remote Sensing Yearbook* [6], *Jane's Spaceflight Directory* [18] and the *Manual of remote sensing* [185]. A particularly useful source of technical data is provided by the alphabetical SATLIS microfiche file, which is part of the RESORS [320] database. Many of the subject-specific textbooks listed in Part II will review sensor systems that are used in their particular field of application. More current information on launches and evolving satellite series may be obtained from the news sections of the newsletters listed in Part II. Of particular interest are specific issues that deal with one sensor; these include the *Columbus Logbook* [342], *Daedalus International* [343], *Dornier ERS-1* [344], *EOSAT Landsat Technical Notes* [352], *Meteosat Image Bulletin* [368], *PST Newsletter* [381], *SPOT Newsletter* [394] and *SPOTLIGHT* [393]. More general newsletter and journal publications that deal with technical satellite developments are *Earth Observation Quarterly* [347], *ESA Bulletin* [353], *NASA News* [370], *NASA Activities* [369], *News from Prospace* [375], *Space* [114] and *Space Markets* [117]. The latter two journal titles are particularly interesting for their coverage of space station development and progress with the Shuttle mission programme.

References

Brookes, A. J. (1975). *Photo reconnaissance: the operational history*. London: Ian Allan.

Colwell, R. N., ed. (1979). *Manual of remote sensing*. Falls Church, Virginia: American Society of Photogrammetry and Remote Sensing.

Curran, P. J. (1985). *Principles of remote sensing*. New York: Longman.

Holz, R. K. (1985). *The surveillant science: remote sensing of the environment*. New York: John Wiley & Sons.

Lillesand, T. M. and Kiefer, R. W. (1979). *Remote sensing and image interpretation*. New York: John Wiley & Sons.

Measures, R. M. (1984). *Laser remote sensing, fundamentals and applications*. New York: John Wiley & Sons.

Sabins, F. F. (1978). *Remote sensing: principles and interpretation*. 2nd edn. Oxford: W. H. Freeman & Co.

2 Who, What, Where ?

The organizational structure of remote sensing is very good and constantly improving. The main reason for such a well defined infrastructure has been the evolution and acceptance of remote sensing from within already established institutional and regional networks. Many of the organizations that are active in remote sensing were formerly photogrammetric in nature; however, they have quickly embraced the new technology and very successfully integrated it into their activities. Indeed, remote sensing has gained security to a degree whereby within the last seven years, two of the most prestigious photogrammetric organizations, the International Society of Photogrammetry and the American Society of Photogrammetry, have appended the phrase 'and Remote Sensing' to their titles.

It is realistic to give only an indication of the ongoing work of the societies and institutions mentioned below, as many of their research projects are in a constant state of flux and new programmes are introduced at regular intervals. It is relatively easy to obtain information on the larger and more well-known organizations, but rather more difficult to acquire equally pertinent data about groups from developing countries. With respect to these less developed countries it has been the policy of this section to concentrate on what are probably the most important elements of their remote sensing programmes: education and training. The names and addresses of all the organizations that are discussed below, and in some cases more complete information, are listed in the Directory section in Part III of this book.

2.1 International

The International Society of Photogrammetry and Remote Sensing (ISPRS) [501] is the largest multinational organization in the world concerned with encouraging co-operation and information exchange in photogrammetry, remote sensing and all related disciplines. Although originally founded in 1907 by Edward Dalezal as the International Society of Photogrammetry (ISP), the name was changed in 1980 as a result of the recognition of the impact of non-photographic sensors in this field. The ISPRS promotes information exchange through the Quadrennial Congress [433] and Symposia [434] of the associated Technical Commissions. Fifteen major meetings have been held since the Vienna Congress in 1913, and the next scheduled congress will be in 1988 at Kyoto, Japan.

Apart from encouraging information exchange through meetings and other means, the ISPRS is concerned with stimulating the formation of new photogrammetric and remote sensing societies, and the co-ordination of photogrammetric and remote sensing research. A great deal of the ISPRS membership's research results are published either in conference proceedings, as volumes of the *International Archives of Photogrammetry*, or in its informative bi-monthly journal *Photogrammetria* [98].

Any country or region that has an interest in the activities of ISPRS may join the society through a single member organization, which is generally that country's national photogrammetry and remote sensing group. Unless they are specially nominated honorary or sustaining members, individuals cannot join the ISPRS, but must be in a member organization or regional association that is affiliated to the society. In addition to the seventy-three member organizations, there are approximately thirty-six sustaining members, which are made up from individuals or organizations providing equipment, services and financial support for photogrammetry, remote sensing and society activities. The final category for society admission is that of an honorary member, which is granted to individuals who have made a special contribution to the society's objectives. A number of honorary awards are also provided by sponsoring members for achievements in certain aspects of photogrammetry and remote sensing.

The central administrative body and policy authority of the ISPRS is the General Assembly, which is composed of one delegate and two advisers from each member organization. Amongst the duties of the General Assembly is the appointment of a council that administers the ISPRS between meetings of the Assembly. The core membership of the various national organizations that are affiliated to the ISPRS constitutes the Congress, which normally meets every four years. Seven Technical Commissions co-ordinate the scientific and technical development of the ISPRS work programme, each being headed by a president and secretary. The Technical Commissions are individually sponsored by a member organization for the four years between Congresses, and as well as organizing a symposium, they appoint working groups on specific aspects of a Commission's work. An excellent summary of the structure and objectives of the ISPRS is provided by a booklet entitled *The ISPRS: Organization and Programs 1984-1988* [247], which in addition to more general information provides the addresses of various Technical Commission and working group members.

The European Space Agency (ESA) [495] was established on 31 May 1975 in order to co-ordinate European space activities. The main ESA initiative of relevance to remote sensing has been the formation of Earthnet, a satellite data receiving, pre-processing and distribution network. However, as far as the current Landsat mission is concerned, the distribution role was taken over by a private consortium on 1 January 1987, following a commercialization policy at ESA member state level.

The Eurimage [721] group, which has now assumed the responsibilities of Earthnet, was formed in 1985 by Hunting Technical Services [725], Telespazio S.p.A. [640], SPOT Image [598], Satimage [694] and the Deutsche Forschungs- und Versuchsanstalt für Luft- und Raumfahrt (DFVLR) [609], with the aim of stimulating industrial and commercial interest in remote sensing. Eurimage will be supplied with Landsat data that have been received and processed by Earthnet, which will continue that part of its activities. The established Earthnet chain of National Points of Contact that have been set up in various countries will be maintained by Eurimage, although there are plans to extend this network to countries which do not have a contact point.

The central headquarters of the ESA are in Paris, but three other main branches exist: the European Space Operations Centre (ESOC) [496] in the Federal Republic of Germany; the European Space Research and Technology Centre (ESTEC) [497] in The Netherlands; and the European Space Research Institute (ESRIN) [498] in Italy. Several sub-organizations exist within the ESA, perhaps the most interesting of which as regards remote sensing is Eumetsat. This group, which was established in 1983, is responsible for operation of the European Meteosat (MOP 1, 2 and 3) programme, including the production and deployment of the satellites. Eumetsat is also likely to be closely involved with certain aspects of the Columbus programme and will be an integral part of the ESA's Earth Observation Preparatory Programme (EOPP). The main concerns of the EOPP, which was initiated in 1985 and should finish during 1990-1991, are ERS-1 and ERS-2, the Polar Orbiting Earth Observation Programmes, second-generation Meteosats and a number of Solid Earth missions.

The ESA, together with the Council of Europe and Commission of the European Communities, funds a dominantly European organization, the European Association of Remote Sensing Laboratories (EARSeL) [494]. EARSeL, which was originally established by six laboratories in 1976, has over 180 member laboratories and some sixty observer laboratories, which are run by a bureau consisting of a chairman, vice chairman, secretary general and treasurer. The membership follows a sound policy: there are no individual members and full membership may be extended only to institutes and laboratories inside Europe that are proposed by two member laboratories, and following a ballot either accepted or rejected by the Annual General Assembly. Observer status is available to those laboratories either outside or inside Europe which do not qualify for full membership. The stated aims of EARSeL are to: promote co-ordination of remote sensing research for peaceful purposes within Europe; facilitate the exchange of ideas and scientific information between participating institutions; identify priority sections of remote sensing research; and to represent members' views to the international (mainly ESA) and national authorities responsible for earth observation systems. The research projects necessary to realize these aims are financed by the sources mentioned above, together with membership dues from each laboratory (which are largely used to maintain EARSeL's structure). The aims are administered through a series of fifteen working groups responsible for research into a wide variety of application areas and education and training.

EARSeL produces two major publications: the annual *EARSeL Directory* [11] and the quarterly *EARSeL News* [349]. These publications are of great value in remote sensing and are more fully described in Part II.

The Organization Européenne d'Etudes Photogrammétriques Expérimentales (European Organization for Experimental Photogrammetric Studies (OEEPE)) [509] was founded in 1953. It is an independent organization, with twelve member countries which are fundamentally concerned with photogrammetry. The group's functions are research into photogrammetry, aerial surveying, topographic mapping and geographical information systems, within an experimental, co-operative framework. The OEEPE is directed by a Steering Committee that meets twice a year, which is in turn supported by a largely administrative Executive Bureau. The Steering Committee is responsible for the appointment of Application Commission Presidents and the Chairpersons of Action Groups and Working Groups. A number of Scientific Commissions have been installed, and are responsible respectively for: Aerial triangulation; Digital elevation models; Restitution; Photogrammetry and cartography; Topographic interpretation and Fundamental problems in photogrammetry.

A newer Commission category, that of Application Commissions, has recently been established for areas such as map revision, digitization of aerial photography, digital photogrammetry, computer-assisted cartography and automatic map digitization. The Working Groups and Action Groups are installed on an irregular basis, as and when a specific topic comes under the scrutiny of the Steering Committee.

The United Nations (UN) has become involved in remote sensing through its Committee on the Peaceful Uses of Outer Space (COPUOS), which has initiated several in-depth remote sensing studies and reviews. On a regional level the UN Economic Commission for Africa (ECA) has become very involved in remote sensing via co-operation with the African Remote Sensing Council (ARSC) [510] and the Regional Remote Sensing Facility [645] in Kenya. The more specialized divisions of the UN, notably the Food and Agriculture Organization (FAO) [499] and World Meteorological Organization (WMO) [503], are concerned with applying remote sensing to the various problems they encounter. The FAO is particularly concerned with education and training, and to this end it annually runs up to ten two- to four-week courses. Around 100 people attend these sessions, which concentrate on the development and management of renewable resources in developing countries. The FAO, together with several other UN agencies, participates in the Global Environment Monitoring System (GEMS), which is under the control of the UN Environment Programme (UNEP). GEMS is concerned with monitoring global climate, air transport of pollutants, renewable resources, the oceans, and human health. An extension of this project, the Global Resource Information Database (GRID), has been initiated as a method of disseminating key environmental data to the user, mainly via large computerized telecommunication networks.

The UN COPUOS liaises with the Committee on Space Research (COSPAR) [604], an international research organization that is affiliated to the International Council of Scientific Unions (ICSU). COSPAR was founded in 1958 by the ICSU, and since then it has been concerned with data exchange on all aspects of satellite and space technology, including remote sensing.

One of the largest and most comprehensive education and training facilities for remote sensing and related disciplines is provided by the International Institute for Aerospace Survey and Earth Sciences (ITC) [500]. Since the ITC was created in 1950 it has had nearly 7,000 graduates from over 145 countries attend its fifty training programmes. Approximately 400 students are provided with training each year, on courses that range from photogrammetry, digital image processing and aerial photography to various remote sensing application-oriented programmes and land use surveys. The qualifications that are awarded for these courses range from Technician and Technologist for the more practical sessions such as photogrammetry and cartography, to postgraduate, MSc and advanced postgraduate degrees for the other courses. The ITC has supported the establishment of regional centres in other countries as part of its technology and education transfer activities. Three centres, which are respectively sited in Colombia [584], India [622] and Nigeria [666], provide courses in Spanish, English, and English and French.

A variety of courses on different aspects of remote sensing are run by the Joint Research Centre (JRC) [502] in Ispra, Italy. The JRC is a European Economic Community (EEC) initiative, with a wide scientific brief that includes remote sensing. The JRC seems to have a particular interest in radar, as well as running courses on synthetic aperture radar (SAR): in co-operation with the ESA it arranged the SAR 580 campaign.

2.2 North America

The U.S. government finalized the takeover of the Landsat programme by the commercial Earth Observation Satellite Company (EOSAT) [741] when the transfer contract was signed on 27 September 1985. EOSAT, a creation of the RCA Corporation and the Hughes Aircraft Company, will undertake a ten-year earth observation programme that is intended to free the company from the substantial government funds it currently receives, and establish it as a viable commercial enterprise.

The foreign ground stations that currently receive Landsat data will still operate under an agreement with EOSAT, which is also planning new stations and station upgrades. Perhaps the most important station will be the new EOSAT data receiving centre in Norman, Oklahoma, which will take over the current NASA Goddard Space Flight Center (GSFC) [762] system. Landsats 4 and 5 are currently being operated through the USGS Earth Resources Observation System (EROS) Data Center [742], but data processing and product generation will be moved to EOSAT headquarters prior to the launch of Landsat-6.

The commercialization of the Landsat programme should ensure data continuity until at least 1995, thus building on the previous fifteen years of data acquisition. EOSAT is not a 'value-added' product vendor itself, although the organization seeks to promote the useful applications of remotely sensed data via its publications, representations at conferences and a widespread network of distributions centres and 'value-added' processors.

SPOT Image, which is more fully described later in this section, also maintains a commercial presence in the U.S.A. The SPOT Image Corporation [789] was founded in 1985 and offers almost the same range of services as its parent company.

It has already been mentioned that EOSAT is very closely affiliated to the EROS Data Center. The National Mapping Division of the U.S. Geological Survey (USGS) [797], which is the parent body of EROS, maintains the Data Center as an official research centre. Amongst its many other functions EROS is a repository for all USGS remotely sensed aircraft and satellite data; it also maintains a comprehensive and specialist library that includes many volumes on remote sensing and related disciplines.

The USGS also maintains the National Cartographic Information Center (NCIC) [773], which is the primary information organization of the National Mapping Division. Millions of items of cartographic and geographic data, including maps, aerial photographs and digital data, from U.S. and non-U.S. sources, are stored at the NCIC. Almost the whole NCIC archive, which is based in Reston, is computerized, thus allowing the rapid retrieval of information from the NCIC itself, and from forty-five state-wide cartographic information centres linked to it. The USGS also runs a very large library system of over one million volumes and 300,000 maps; most of this system is open to the public, or accessible via interlibrary loan.

The American Society for Photogrammetry and Remote Sensing (ASPRS) [740] is the largest scientific society in the U.S.A. that is directly concerned with remote sensing. Originally founded in 1934 as the American Society of Photogrammetry (ASP), the name change occurred in early 1985, in recognition of the contributions being made to photogrammetry and related disciplines by remote sensing. The membership of the ASPRS runs to some 8,000 individuals, and the principal aims of the Society are to advance knowledge in all areas of photogrammetry and remote sensing.

The ASPRS has very close links with the American Congress on Surveying and Mapping (ACSM) [745], a society composed of three separate organizations: the American Cartographic Association (ACA), the National Society of Professional Surveyors (NSPS) and the American Association for Geodetic Surveying (AAGS). The ASPRS and ACSM hold two joint meetings each year; these differ in the technical programmes presented, but all other aspects of organization are shared by the two societies.

The ASPRS is perhaps best known, and revered, for its numerous publications, which include the *Manual of remote sensing* [185] and what is probably the most prestigious of all remote sensing journals, *Photogrammetric Engineering and Remote Sensing* [99]. As the leading National Society for photogrammetry and remote sensing, the ASPRS is the U.S. national member of the ISPRS.

One could not discuss U.S. involvement in remote sensing and related disciplines without mentioning several of the largest and most beneficial sponsors of space activity: the National Aeronautics and Space Administration (NASA) [772], the National Oceanic and Atmospheric Administration (NOAA) [774] and support services such as the National Environmental Satellite Data Information Service (NESDIS). For decades NASA and NOAA have been responsible for the funding of all major earth observation missions, and it is only recently that they have relinquished their grip on these programmes, following commercialization policies at government level. NOAA is still maintaining an active and very successful role in the development and maintenance of meteorological satellites in the TIROS-N and GOES series, although it is reliant on NASA for the launch of these missions. NASA's current involvement in remote sensing is mainly in the support and research modes, with active research being concentrated on the development of an International Space Station. Unfortunately this agency, and the aerospace industry as a whole, has suffered from delays in satellite launch capability, which culminated in the tragic loss of the Space Shuttle Challenger, and accidents with NASA's Expendable Launch Vehicle (ELV) programme. It is unlikely that the Space Shuttle will be flown during 1987 or 1988, and although NASA has reached Memorandums of Understanding with several countries involved in space station research, its position as a leader in launch technology will now be challenged.

NASA maintains a number of official research centres, of which one of the most interesting is the designated NASA Industrial Application Center, the Technology Application Center (TAC) [743] in New Mexico. The TAC is particularly concerned with remote sensing, and is heavily involved in the provision of data services, education and technology transfer in relation to this commitment. The TAC also acts as an archive for satellite and aircraft imagery in its role as a centre for the NCIC, and as a repository for NASA and other government documents, with in-house online access to one of the most complete document libraries on remote sensing in the U.S.

A great number of other organizations that have a direct involvement with remote sensing exist in the U.S.A. Many of these are academic or predominantly educational institutions, and as such are far too numerous to list here; however, several are outstanding, including the Laboratory for Applications of Remote Sensing (LARS) [768], the Environmental Research Institute of Michigan (ERIM) [759] and the Geosat Committee Inc. [761], which is affiliated to NOAA. A similarly large number of commercial and industrial companies are involved in remote sensing, such as E. Coyote Enterprises [758], DIPIX [755], Earthsat [756] and Daedalus [754]. A useful list that analyses private sector involvement in remote sensing has been compiled by Carter (1984). This bibliography outlines commercial remote sensing concerns and provides a preliminary listing of 150 companies, which gives their name, location and the type of products and services that they can offer.

Other institutions and societies, although not always directly involved in remote sensing, nevertheless publish a great deal of material in related areas. The most notable of these is the IEEE International Geoscience and Remote Sensing Society (IGARSS) [763], which has its own journal, *IEEE Transactions on Geoscience and Remote Sensing* [61], and regularly hosts meetings on various specialist topics in remote sensing. The other main groups involved in this type of activity include the Society for Photo-Optical Instrumentation Engineers (SPIE) [787], the Society of Photographic Scientists and Engineers (SPSE) [788] and the American Society of Civil Engineers (ASCE) [748].

The focus for remote sensing in Canada is the Canada Centre for Remote Sensing (CCRS) [539], located in Ottawa, Ontario. Established in 1972 as the central element in Canada's national remote sensing programme, the CCRS is very active in all fields of remote sensing research and development. The co-ordinating body and advisory group on policy and financial matters is the Interagency Committee on Remote Sensing, which has members from government agencies and departments. The Canadian Advisory Committee on Remote Sensing (CACRS) also operates from the CCRS, and consists of chairpersons from twelve application-oriented working groups, representatives from the provinces and territories, and heads of speciality centres and organizations. Provincial centres and co-ordinators for remote sensing activities have been established right across Canada: in British Columbia [562], Alberta [547], Saskatchewan [569], Manitoba [560], Ontario [540], Quebec [570], Maritimes [561], New Brunswick [565], Nova Scotia [566], Prince Edward Island [567], Newfoundland [563], Yukon [568] and the Northwest Territories [555].

One of the most useful initiatives of the CCRS, and one that provides a much-needed service to the rest of the world's remote sensing community, is the setting-up of the Remote Sensing Online Retrieval System (RESORS) [320]. This online database not only houses a massive number of bibliographic citations on remote sensing, but also provides lists of available slide sets, acronym indexes, and reviews and previews of the world's progress with spaceborne remote sensing experiments.

The CCRS has also played an important role in the establishment of a Canadian Remote Sensing Training Institute (CRSTI) [551] in Ottawa. The Institute has a wide brief that includes the documentation of Canadian remote sensing activities, maintenance of up-to-date information on training needs, and promotion of the training capabilities available within Canada for remote sensing and related disciplines. The non-profit-making Canadian Remote Sensing Society [550] is another Canadian organization that is largely concerned with education and training. The society regularly sponsors the Canadian Symposium on Remote Sensing [420] and also publishes the bi-annual *Canadian Journal of Remote Sensing* [36] as part of its educational activities.

2.3 Western Europe

The main centre for British remote sensing activities is the National Remote Sensing (NRSC) [703] at Farnborough, funded under the British National Space Centre (BNSC) [715] programme. Although it has been operational for some years, the official opening was delayed until 22 September 1986. The NRSC is interested in all aspects of remote sensing, including user education, which is normally carried out with the provision of short courses at Silsoe College [734], one of its regional centres.

Commercial interest in remote sensing applications is largely covered by the British Association of Remote Sensing Companies (BARSC) [710]. BARSC, a conglomerate of sixteen British companies involved in remote sensing, was formed in spring 1986 with the aim of establishing a private-sector body representative of U.K. remote sensing interests. Some of the companies involved in BARSC (Hunting Technical Services Ltd. [725], General Technology Systems Ltd. [723], Clyde Surveys Ltd. [719] and Nigel Press Associates Ltd. [729]) already have considerable experience in remote sensing and should provide a solid basis for the operation of BARSC in the international marketplace.

Two British learned societies are respectively concerned with photogrammetry and remote sensing: the Photogrammetric Society [732] and the Remote Sensing Society (RSS) [733]. The Photogrammetric Society promotes a number of meetings, often co-sponsored by the RSS, and publishes the journal, *Photogrammetric Record* [101]. The RSS was founded in 1974 to bring together professionals engaged in remote sensing and to provide a forum for discussion of their work. The results of ongoing research are aired at the annual conference of the RSS and in its serial publication *Remote Sensing Society's News and Letters* [387].

Several U.K. universities are involved in remote sensing research and consultancy. Some of these, such as Dundee University [716], Imperial College [718] and Silsoe College [734], run one-year MSc courses. Both these and other academic institutions, like the Remote Sensing Unit (RSU) [704] at Aston University, offer postgraduate research programmes. In addition to formal research qualifications, the RSU offers specialized education and training courses for foreign and domestic students in a wide variety of application areas.

The Centre National d'Études Spatiales (CNES) [603], which was created in 1961, is a commercially oriented enterprise that is administered by the French government. This French space agency is heavily committed to remote sensing activities, and its major earth observation initiative to date has been the SPOT satellite system. CNES funded the development of SPOT and continues to support its operation. In keeping with the CNES doctrine of profitability, a figure approaching $41 million from private and government sources has been used to establish the distribution company SPOT Image [598], which will handle data sales and distribution.

CNES considers that it will recoup the total cost of the SPOT programme at a rate of $70-100 million a year. This supposition may appear somewhat optimistic as this revenue will be almost entirely from data sales; however, it is quite realistic, given the complex worldwide network of forty-seven SPOT data distributors that has already been established. SPOT is competitive and very professional; its selling policies are a combination of aggressive marketing backed up by an efficient business service that includes online computerized access to information concerning image data.

CNES is in close liaison with the Institut Géographique National (IGN) [597], a research centre that amongst its other activities has interests in aerial surveying, photogrammetry, cartography and remote sensing. The IGN is closely allied to the preparation of CNES earth observation missions, and is at the forefront of development in methods of exploiting such data for cartographic purposes.

The CNES and IGN established the Groupement pour le Développement de la Télédétection Aérospatiale (GDTA) in 1973. GDTA has three members, the Bureau pour le Développement de la Production Agricole (BDPA) [601], the Bureau de Recherches Géologiques et Minières (BRGM) [600] and the Institut Français du Pétrole (IFP) [606], which are concerned with education and training, the development of remote sensing services, and commercialization. The GDTA is particularly active in training and offers a variety of courses for thirty-four weeks, producing about 150 graduates each year. Distribution of Landsat products is also currently handled by the GDTA, which has recently signed a contract with the ESA for the processing of raw data from ERS-1.

The French Prospace [599] association is a non-profit organization consisting of over forty firms and groups that are active in the aerospace industry. Prospace was founded in 1974 by CNES and aims to keep space specialists worldwide informed of French space activity, while at the same time establishing international links between these specialist groups. Prospace is actively involved in the SPOT and Ariane programmes, and should be contacted by any group or individual interested in these missions; all Prospace services are free of charge.

The largest research establishment dealing with the aerospace sciences in the Federal Republic of Germany is the Deutsche Forschungs- und Versuchsanstalt für Luft- und Raumfahrt (DFVLR) [609]. The DFVLR has several centres, and although these are largely concerned with aerospace research, they offer some education and training opportunities connected with remote sensing, with research into optical and microwave sensor processing technology. The Remote Sensing Research Department is located at Oberpfaffenhofen and has been heavily involved in the Metric Camera and Modular Opto-electronic Scanner (MOMS) experiments that were flown on the Space Shuttle. DFVLR is committed to the ERS-1 programme and is still involved in processing Seasat data in connection with the development of its own Synthetic Aperture Radar (SAR). The DFVLR liaises with the Institute for Photogrammetry at the University of Hannover, the home of the German Society for Photogrammetry and Remote Sensing [613], which is the ISPRS representative in Germany.

There are three branches of the Institut für Angewandte Geodäsie (IFAG) [610], but only the Cartographic and Photogrammetric Research Divisions hold any real remote sensing interest. The IFAG's researchers are interested in applied research that has a practical end-use, and in this connection they are involved with digital image processing, geometric corrections, automatic data processing and a wide variety of applied cartographic techniques. The IFAG has a very large library of nearly 60,000 volumes; interestingly, the library houses a translation bureau for foreign-language publications.

The main centre for remote sensing activity in Italy, apart from the ESA ESRIN [498] facility, is the state-owned industrial Telespazio S.p.A. [640] complex in Rome. Telespazio is a major concern, which liaises with the ESA and Consiglio Nazionale delle Richerche (CNR) [636], and is heavily involved with satellite communications; indeed, the company is responsible for operation of the Landsat receiving ground station at Fucino. Telespazio is the nominated vendor of SPOT products in Italy, and will be receiving and processing data from the ERS-1 satellite when it becomes available.

The Società Italiana di Fotogrammetria e Topografia [639] has been active in Italian remote sensing for some time. It has been joined more recently by the Italian Remote Sensing Association (AIT), which operates from Milan. The AIT's aims are to represent domestic remote sensing interests in the international community and stimulate awareness of the Italian remote sensing scene.

At the end of 1983 the Netherlands Cabinet Council established a well structured National Remote Sensing Programme (NRSP), with an initial five-year development plan. One of the first steps in this plan was the reorganization of the Netherlands Remote Sensing Board (BCRS) [660], which took place in May 1986. The BCRS Programme Bureau, a team of three, are responsible for implementation of the NRSP, which involves close liaison with the ESA, preparation for ERS-1, application-oriented work, and participation in industrial and commercial activities.

The National Aerospace Laboratory (NLR) [658] is the second organization involved in remote sensing in the Netherlands. The NLR has fairly close ties with the BRSC and has over fifteen years of experience in remote sensing. NLR is both the Eurimage and SPOT satellite data distributor for the Netherlands and also operates several aircraft of its own for the acquisition of a range of remote sensing data.

The Société Belge de Photogrammétrie, de Télédétection et de Cartographie (SBPT) [529] is the oldest-established remote sensing organization in Belgium, and the national member of the ISPRS. A more recent initiative in Belgium has been the installation of a remote sensing unit within the Belgian Science Policy Office (SPO) [528]. A National Remote Sensing Programme (NRSP) was initiated on 1 August 1986, with the involvement of seventeen research laboratories. The SPO was involved in financing the SPOT project and is the distribution agent for SPOT in Belgium. Consolidation of the Belgian NRSP is ongoing, and the Science Policy Office should be contacted for further details.

The Belgian Institut Géographique National (IGN) [530] is also involved in remote sensing, photogrammetry and computer-assisted cartography. Although administered by the Ministry of National Defence, the IGN is a Study Group of the Parliamentary Assembly of the Council of Europe and is involved in information exchange and co-operative work at a number of levels.

There are a number of other European centres that are involved in remote sensing. The Instituto Geográfico Nacional (IGN) [687] in Spain and the Instituto Geográfico e Cadastral (IGC) [680] in Portugal use remote sensing data, but primarily in photogrammetric and cartographic applications. Sweden has a commercial/government interest through the Swedish Space Corporation (SSC) [692], which runs short education and training courses and maintains a Landsat and SPOT receiving station at ESRANGE [693]. The Department of Photogrammetry [501] at the Royal Institute of Technology in Stockholm is well regarded, and at the time of writing it is the current contact point for the Secretary General of the ISPRS. The Geodeettinen Laitos (Geodetic Institute) [595] of Finland is involved in remote sensing, although primarily on the application of photogrammetric methods in geodesy and surveying.

2.4 Eastern Europe and the U.S.S.R.

Poland has a fairly well established remote sensing community focussed on two main centres, the Polish Remote Sensing Centre at the Instytut Geodezji i Kartografii [676] in Warsaw, and the Photogrammetric Section [674] of Krakow's University of Mining and Metallurgy. The former centre is engaged in research into photogrammetry, development of remote sensing methods and applications, image interpretation and thematic mapping using Landsat and Salyut imagery. Although involved in this sort of work, the University is more concerned with education and training, and to this end it runs several courses and occasional meetings that are organized in conjunction with the Polish Society for Photogrammetry and Remote Sensing. Short to medium-term courses of up to three months' duration are also held at the Polish Remote Sensing Centre, with the language of instruction being either Russian or English. Similar training programmes, again largely using domestically derived national imagery for application programmes, are run by the Czechoslovakian Remote Sensing Centre [587] in Prague.

Földmérési Intézet (FÖMI) [618] is a government-sponsored geodesy and cartography institute situated in Budapest. The Remote Sensing Center (RSC) is primarily interested in digital and analogue satellite and airborne image analysis, although it is also concerned with modern photogrammetric surveying research. A large library is housed at FÖMI and there are substantial collections of satellite data and aerial photographs, some of which are used in the occasional training courses run by the RSC. FÖMI, which distributes satellite data for Eurimage [721] and SPOT Image [598], is involved in the Soviet-bloc Intercosmos programme.

The International Co-operation in Research and Uses of Outer Space (Intercosmos) Council was formed in 1967 under the aegis of the Academy of Sciences of the U.S.S.R. [806]. Ten countries (Bulgaria, Czechoslovakia, Cuba, the German Democratic Republic, Hungary, Mongolia, Poland, Romania, the Soviet Union and Vietnam) are involved in the programme, which undertakes a variety of space experiments in Soviet-sponsored rocket launches. Remote sensing has been part of Intercosmos's activity for over eleven years and the Academy of

Sciences has a number of remote sensing research groups, which are mainly concerned with the interpretation of images from the Meteor meteorological satellite series.

The education and training aspects of remote sensing in the U.S.S.R. are catered for by the Moscow Institute of Engineers for Geodesy, Aerial Surveying and Cartography (MIIAiK) [808] in Moscow. Some 300 students are currently enrolled on the six-year course offered at MIIAiK and tuition is both practical and theoretical, including lengthy modules on aerial photography, air surveying, cartography, geodesy, photogrammetry and photointerpretation. A number of fellowships are available under the UN Programme on Space Applications, and there are facilities for Russian-language instruction if a participant is not a Russian national.

2.5 Asia and the East Pacific

The Department of Photogrammetry and Remote Sensing [580] at the Wuhan Technical University of Surveying and Mapping in China provides education and training at a number of levels. The courses, which have variable amounts of photogrammetry and remote sensing, lead either to BSc or MSc qualifications. The national programme for remote sensing in China was established by the State Scientific and Technical Commission (SSTC) in 1981. There are several SSTC-administered centres, most of which, like the China National Committee on Remote Sensing [582], fall under the auspices of the Academia Sinica (AS). Application-oriented problem-solving research is performed at the Scientific Research Institute of Surveying and Mapping (SRIS) [579] based in Beijing. Increasingly, China appears to be launching reconnaissance and earth observation missions; the Chinese are also competing in the international launch market by offering commercial satellite launch facilities via their ELV's, which are based at Chengdu.

The Indian Space Research Organization (ISRO) [623] is a nodal organization that co-ordinates activities related to remote sensing technology, applications and training and education programmes in India. The National Remote Sensing Agency (NRSA) [620] operates under the aegis of the ISRO and its principal activities are receiving and processing image data, and running special short courses on all aspects of remote sensing. A great deal of the NRSA's educationally oriented duties are performed at an offshoot of NRSA, the Indian Institute of Remote Sensing (IIRS) [622] at Dehra Dun. The IIRS, which is closely affiliated to the ITC [500], runs diploma courses of up to ten months' duration in all aspects of practical remote sensing. There are facilities for digital image processing at the Institute as well as extensive collections of satellite and aircraft imagery, much of which is used for teaching purposes. The Centre of Studies in Resources Engineering (CSRE) [621] at the Indian Institute of Technology in Bombay offers both short and long training courses in remote sensing for national and foreign scientists and engineers. The research activities at this centre include image processing, microwave remote sensing, water resources and mineral resources.

More specialized training is offered at the Space Applications Centre (SAC) [624], an ISRO affiliate. The SAC, which is involved in the Indian Remote Sensing Project, concentrates on training in specific applications, including agriculture, coastal processes, crop stress monitoring and yield prediction, forestry, water resources, ocean resources, digital image processing and geographic information systems. India is also active in earth observation and currently has one operational Bhaskara satellite (launched by the U.S.S.R.), with a second remote sensing mission, IRS-1A, due for launch in the 1987/88 time-frame.

The promotion and co-ordination of remote sensing technology, applications and activities in Pakistan is handled by the Remote Sensing Applications Centre (RESACENT) [670] of the Space and Upper Atmosphere Research Commission (SUPARCO). A considerable collection of satellite data is archived at SUPARCO, which is currently negotiating Landsat and SPOT reception and distribution rights for Pakistan, following the completion of its own satellite data receiving station in 1986/87.

The Remote Sensing Technology Centre (RESTEC) [641] of Japan is a commercial foundation sponsored by several large concerns, including the National Space Development Agency (NASDA) [643]. RESTEC is fundamentally involved in research and education, together with satellite data acquisition, processing, analysis and image distribution. NASDA is currently negotiating reception and distribution rights with SPOT Image; RESTEC already receives and distributes Landsat data. The major remote sensing initiative of this country is the forthcoming (1987) launch of an ocean monitoring satellite, MOS-1. The Japan Society of Photogrammetry and Remote Sensing [642] is the ISPRS representative in Japan, and will host the next ISPRS Quadrennial Congress at Kyoto in 1988.

The Thailand Remote Sensing Centre (TRSC) [700] was established in 1972 by the National Remote Sensing Co-ordinating Committee, which is part of the Thai National Research Council (NRC). The major focus of Thai research is on agriculture; particularly crop monitoring, crop yield forecasting and forestry. A very substantial archive of satellite data is maintained at the TRSC, which already distributes both Landsat MSS (which it also receives) and SPOT imagery, and is hoping to upgrade its station's capability to receive Landsat TM and SPOT data.

The Asian Institute of Technology (AIT) set up the Asian Regional Remote Sensing Training Centre (ARRSTC) [701] in Bangkok as an attempt to co-ordinate technology transfer within the Asian and Pacific region. The ARRSTC is dominantly concerned with education and runs three courses a year for anything between 100 and 150 students. Entrants, who are selected from remote sensing centres within the region, are highly trained in modern computerized techniques of digital image analysis, as well as more traditional aspects of remote sensing.

There are a number of other remote-sensing-related organizations in this region, and centres are located in Bangladesh [527], Malaysia [648], Nepal [657], the Philippines [673], Taiwan [699] and Sri Lanka [689]. One of the most active of these is the Bangladesh Space Research and Remote Sensing Organization (SPARSSO), which is involved in practical remote sensing training in application-oriented subjects.

Most of the countries and organizations that have been discussed above belong to the Asian Association on Remote Sensing (AARS) [504], together with countries such as Korea [646] and Vietnam [814]. The AARS was founded in 1981 in an effort to promote co-operation and information exchange between remote sensing organizations, and much of this work is done at the annual conference, which is hosted by a different member country each year.

2.6 Australia and New Zealand

Australia is ideally suited to the exploitation of its land resources by remote sensing, as its high area to population ratio is unable to economically support exploration by conventional means. Much of the work done on remote sensing in Australia is co-ordinated by the Commonwealth Scientific and Industrial Research Organization (CSIRO) [518], which has several research outlets. Landsat MSS data are received at the Australian Landsat Station (ALS) [515] in Belconnen, which publishes the *Australian Landsat Station Newsletter* [334], an informative guide to what is happening in Australian remote sensing. The headquarters of the Division of National Mapping (NATMAP) [520], which is primarily a cartographic organization, are also situated at the Belconnen site.

The Australian Mineral Foundation (AMF) [513] is mainly concerned with remote sensing education and training for professionals in the mining and petroleum industries. The thorough courses offered at the AMF are backed up by excellent library facilities, including computerized information services which generate several bibliographic publications of direct relevance to remote sensing. The Remote Sensing Association of Australia (RSAA) [514], housed in the Centre for Remote Sensing (CRS) at the University of New South Wales (UNSW), is involved in education to the extent of hosting the triennial Australasian Remote Sensing Conference. The CRS also offers postgraduate research programmes in remote sensing and has excellent facilities; the central library of the UNSW is a nominated repository for NASA publications. Research and development, together with education, is undertaken by the South Australian Institute of Technology (AIT) [522]. The AIT only has a small research staff, but is involved in agricultural and environmental resources monitoring.

In New Zealand the central organization for research into different types of remotely sensed data and the provision and distribution of related products is the Division of Information Technology (DIT) at the Department of Scientific and Industrial Research (DSIR) [661]. The DSIR is not an educational institution, although a visiting scientists programme is operated for researchers interested in remote-sensing-related work undertaken at the DIT. More formal opportunities for education exist at the University of Auckland, where MA degrees are offered with a large remote sensing component. The Physics and Engineering Laboratory (PEL) [664] at the same institution is more research oriented, but does conduct some short-term training seminars. Applied research projects into the practical uses of remotely sensed data for soil and water conservation are carried out by the Remote Sensing Group (RSG) [663] of the Soil Conservation Service. The RSG is involved in digital image analysis via a co-operative agreement with the DSIR, and is one of the groups participating in the SPOT data preliminary analysis programme.

2.7 South and Central America

The largest centre for remote sensing within South America is the Instituto de Pesquisas Espaciais (INPE) [532], an official research organization affiliated to the Conselho Nacional de Desenvolvimento Científico e Technológico. INPE has recently become very involved in space activities and is developing its own satellite launch capability. The INPE receives Landsat data at the Cuibá station, and these are distributed throughout South America. The Institute is also currently negotiating for the same deal on SPOT imagery with SPOT Image. Short courses are run for professionals at INPE; these are generally for Latin American nationals, with the language of instruction being either Portuguese or Spanish. Three-month specialist courses, together with both undergraduate and postgraduate training programmes, are held at the Instituto de Geociências e Ciências Exatas (UNESP) [533], which is the second largest establishment for remote sensing, after INPE, in Brazil.

The Centro de Levantamientos de Recurses Naturales por Sensores Remotos (CLIRSEN) [591] in Ecuador is a government-financed organization that provides training and technological expertise to other groups. CLIRSEN was instrumental in the establishment of the Society of Latin American Specialists in Remote Sensing (SELPER), an organization concerned with co-operation and technology transfer within Latin America.

The Centro Interamericano de Fotointerpretación (CIAF) [584] was established in 1968 as a government and partly commercial institution. CIAF is very heavily involved in user education, and like the IIRS [622] in India, is closely affiliated to the ITC [500]. Numerous short and specialized courses are offered, including a series of postgraduate training programmes in forestry and ecology, geology, soils science, regional survey and civil engineering; detailed booklets describing these courses are available from the CIAF. Some fellowship funding is forthcoming from the Organization of American States (OAS), the Dutch government, and the Interamerican Development Bank (IDB); via financial aid, the latter two organizations are helping the CIAF to implement a remote sensing programme.

The remote sensing programme in Chile comes under the aegis of the military, particularly the Sección Sensores Remotos of the Instituto Geográfico Militar [576]. The other group of interest in this country, and the nominated distributor of SPOT data, is the Servicio Aerofotogramétrico [578], an Air Force-affiliated organization. Chile does not appear to be using remotely sensed data from satellites very extensively; however, aerial photography, photogrammetry and cartography are commonly used techniques.

Although several other Latin American countries, such as Bolivia [531], Mexico [653, 654, 655], Nicaragua [665], Peru [672] and Venezuela [812, 813], have an interest in remote sensing, only Argentina and Panama appear to be fully exploiting the technology. The Comisión Nacional de Investigaciones Espaciales (CNIE) [512] in Buenos Aires distributes SPOT data, and receives, processes and markets Landsat MSS imagery through its own ground station. CNIE also collaborates in educational matters via the UN and other national agencies, although the centre is not educationally oriented. The Defense Mapping Agency (DMA) [671] in Panama does run a number of courses, as well as providing training that is dominantly cartographic in nature, although the courses do feature some remote sensing and photogrammetry. The DMA also houses a very good library, probably the best in Latin America outside Brazil, that contains microfiche Landsat browse files and numerous remote sensing articles on fiche.

2.8 Africa and the Middle East

The UN ECA is involved in collaborative work with the African Remote Sensing Council (ARSC) [510], with the aim of establishing five regional remote sensing centres in Africa. The ARSC has over twenty member countries and the centres have been set up with the aim of co-ordinating their remote sensing activities. Perhaps the best-known centre is the Regional Remote Sensing Facility (RRSF) [645] in Kenya. The RRSF provides training to established professionals for up to six months; the emphasis of these courses is practicality, and the coverage includes natural and man-made disasters, highway engineering and pest monitoring. The RRSF is currently seeking affiliation to the ITC [500] educational programme, an alliance that has been granted to several other organizations previously mentioned in this section. The Centre Régional de Télédétection de Ouagadougou (CRTO) [538] in Burkina Faso and the Programme d'Études des Ressources Terrestres par Satellite (ERTS) [817] in Zaire are also involved in the ARSC programme, together with the Remote Sensing Centre [592] in Egypt and the ECA Regional Centre for Training in Aerial Surveys (RECTAS) [666] in Nigeria. The latter two centres were formerly national remote sensing institutes that are being extensively re-organized for their roles as regional representatives. Danz Surveys and Consultants [667] distribute SPOT data in Africa, along with organizations in Malawi [650] and Tunisia [702].

The government research centres in Madagascar [647] and Zaire [816] are largely concerned with photogrammetry and aerial surveying. Although heavily involved in photogrammetry and mapping, the Direcçâo Nacional de Geografia e Cadastro (DINAGECA) [656] in Mozambique regularly utilizes aerial photography (obtained under contract) in its studies, and the Geological Survey Department [818] in Zambia frequently uses remote sensing and photogeological techniques for regional geological investigations. The extent to which remote sensing is being used in African countries is illustrated by the Zimbabwean National Committee for Remote Sensing [819], which flies aerial photographic cover of the whole country every five years, and is currently becoming involved in digital processing of satellite image data.

The Satellite Remote Sensing Centre (SRSC) [684] of the National Institute for Telecommunications Research is the focal point for remote sensing within the

Republic of South Africa. The SRSC distributes both Landsat and SPOT images, receiving the MSS data at its own station. A considerable archive of over 38,000 Landsat 1, 2, 3, 4 and 5 scenes is maintained at the SRSC, which also provides specialist digital image analysis equipment and related training facilities.

A great deal of remote-sensing-related work is undertaken in Middle-Eastern countries; however, a substantial amount of this is oriented towards military applications and is therefore classified. The Remote Sensing Department [630] in Iraq is a government-funded civilian organization concerned with remote sensing research, development and applications. The Department has both image processing and photographic laboratories, to which a few trainees are admitted each year. Iran also supports a Remote Sensing Center [629] that is involved in application-oriented research and the administration of national remote sensing interests. Jordan houses a Remote Sensing Center (RSC) [644] in its National Geographic Center, which is located in Amman. Cartography and geodesy are amongst the RSC's other interests, which also include short training courses in the principles and applications of remote sensing.

Education and training are fundamental aspects of the programme run by the National Remote Sensing Center (NRSC) [698] in Syria. Four- to eight-week courses on the applications of remote sensing have recently been introduced at the NRSC, which also offers a variety of research topics. One other Syrian organization, the Military Survey Department [697], also based in Damascus, is involved in aerial photography, photogrammetry and surveying. The Interdisciplinary Center for Technology Advanced Forecasting (ICTAF) [635] is perhaps the most active information group in the Middle East. ICTAF is particularly involved in space and remote sensing research and development, including online database networks. This interest is reflected by two very comprehensive and useful serial issues, the *Israel Remote Sensing Information Bulletin* [363] and *Space Research and Technical Information Bulletin* [391].

Landsat data for the Middle East region are distributed by the King Abdulaziz City for Science and Technology [682] in Saudi Arabia. This Institute, which receives and processes Landsat MSS data at its own station, is negotiating for the right to receive and distribute SPOT images.

Reference

Carter, W. D. (1984). The private sector: a global pool of talent for remote sensing training and programme support. *Advances in Space Research* 4 (11): 49-57

3 The Remote Sensing Literature

3.1 General Characteristics

The literature of remote sensing is very diverse, but is to some extent constrained by its evolution within long-established subjects such as geology and photogrammetry. This dissemination has resulted in a great deal of the early literature on remote sensing being somewhat 'hidden'; however, although remote sensing is still scattered to some degree, it has more recently attained a certain recognition, and literature trends are more discernible. The increased awareness of remote sensing is most strongly indicated in the periodical literature, where it first appeared in serials concerned with photogrammetry, aerial surveying and geodesy. Many of these journals are now fundamentally involved with remote sensing and merely pay lip service to their former focus of attention. One problem with these publications is that they continue to use titles that are slightly misleading, as they do not provide a totally accurate idea of content and scope. Thankfully this problem is being reduced, via changes in journal name, and the appearance of new subject-specific periodicals.

With regard to book literature, remote sensing is adequately represented, although texts on more specialized aspects of the science are being needed increasingly. Many of the books that are presently available tend to divert themselves from the main topic of interest and indulge in lengthy introductions to the physical basis of remote sensing and available sensor systems. This can be useful to the novice, but several excellent introductory texts are currently available, which makes the duplication of this work seem pointless.

3.2 Classification in Remote Sensing

The Dewey Decimal Classification ('Dewey') and Library of Congress Classification seem to cater for remote sensing and related disciplines quite well, although the Universal Decimal Classification (UDC) is appropriate in places. Dewey certainly provides ready accommodation for remote sensing at 621.367, the Technological photography and photo-optics sub-section of Applied Physics (621). Remote sensing technology is classed as 621.367 8, with books such as *Principles of remote sensing* [186] (621.367 8 19) and *Remote sensing: the quantitative approach* [221] (621.367 8 18) slotting into the classification. Sensor technology tends to be covered at Electronic Engineering (621.381) by Photo-electric and photo-electronic devices (621.381 542). A great deal of remote sensing material is placed at 526, 550 and 551, respectively Mathematical Geography, Sciences of the Earth and Other Worlds, and Geology, Meteorology and General Hydrology. It is here that remote sensing in oceanography, hydrology and meteorology is classified, with *Satellite sensing of a cloudy atmosphere: observing the third planet* [156] (551.4 18) being an example. It is perhaps in this area that remote sensing is at its vaguest in the Dewey classification, as application-oriented books such as *Remote sensing in geology* [181] (551.028 19) are generally the norm.

Photogrammetry and cartography occurs under the Surveying (526.9) sub-section in the Mathematical Geography (526) section. Photogrammetry is represented at 526.982, and Aerial and space surveying at 526.982 3. The Military application of photogrammetry may be searched for at the 623.72 classmark, with Reconnaissance aircraft being placed at 623.746 7. Dewey also specifies Special kinds of photography at 778.3; remote sensing is represented here by 778.35, Aerial and space photography (including interpretation).

Under UDC a similar classification to Dewey is observed, although the author cannot find remote sensing included as such. Some of the Dewey examples above are similar in the UDC: Reconnaissance aircraft and Observation aircraft are at 623.746.2, Aerial and space photography at 778.35, and Meteorology at 551.5; with 551.501.82 being Meteorological satellites, in comparison with 551.635.4 under Dewey. Geodesy, Surveying, Photogrammetry and Cartography are at the 528 classmark, which differs from Dewey. The principal divisions of interest here are Photogrammetry at 528.7 and Interpretation of aerial photographs, 528.77.

The Library of Congress Classification deals with remote sensing in a thorough, if dated, manner. Remote sensing is defined as a sub-heading under both Geology (QE 33.2.R4) and Ecology (QH 541.15.R4), whilst Aerial photography in geology is present at QE 33.2.A3. In fact, aerial and space photography, and photogrammetry, are fairly well established in this classification; Photographs from space is at the QB 637 classmark under Astronomy; the subheadings of Aerial photography and Photographic interpretation occur at TR 810; and Photogrammetry is accommodated by TR 693. Photography in surveying and aerial surveys are respectively represented under Engineering (TA 593) and Highway Engineering (TE 209), Aerial surveying being present at classmark TE 209.5 under the latter heading.

Meteorological satellites are accommodated under the unlikely heading of Motor vehicles at TL 798.M4, with Weather satellites being located at the QC 879.35 mark (the Physics heading, which also accommodates QC 829.55.84, Photography including automatic picture receiving and transmission). Radar and scanning systems are classed with the TK prefix: Electrical Engineering. Electronics. Mechanical Engineering. TK 6595 and TK 788.2 respectively identify Special applications of radar and Scanning systems.

Military aspects of remote sensing are confined to the UG, Airforces. Air Warfare and VG, Minor Services of Navies headings. The former area incorporates reconnaissance planes at UG 1242.R4 and Aerial reconnaissance and Reconnaissance satellites at classmark UG 760-765. Photographic interpretation is apparently a Naval exercise at VG 1020.

Both the Dewey and Library of Congress classifications are quite limited in the way that they cover remote sensing. Often there are no specific numbers for the various sub-divisions of the science, meaning that a wide variety of documents are placed under a single classmark. These generalized numbers may work to some extent for large libraries; however, more specialized subject-specific document stores suffer with this approach. The situation is exacerbated by the lack of UDC control, as this more specific and subdivided classification scheme, which is often used for libraries that specialize in one area, does not include remote sensing.

3.3 Research Directories

Directories and monographs that list and review general sources of scientific information can provide very useful records of organizational scope and activity. A number of these publications have been compiled specifically for remote sensing, although others that are not dedicated to this science can also be very useful.

The most recently published major international source of directory-type information concerning the status of remote sensing research, education and training is the *Remote Sensing Yearbook (1986)* [6]. Containing more than 500 entries on a variety of countries, together with invited papers on the status of remote sensing in some of these areas, it provides an excellent overview of industrial, commercial and academic organizations involved in remote sensing. A smaller and less comprehensive directory of aerospace organizations offering education and training facilities was compiled by the United Nations Department of Political and Security Council Affairs in 1986. The publication, *Education, Training, Research and Fellowship Opportunities in Space Science and Technology and Its Applications* [24], contains a section (pp. 67-161) specific to remote sensing, which lists information on personnel, facilities, courses and tuition fees. While the guide is fairly basic and not fully comprehensive, it does give a very useful indication of the scope for training in various UN countries.

Dominantly European in outlook and content, the *EARSeL Directory* [11] has been published and supplemented annually for nine years. The *EARSeL Directory* lists commercial and industrial European members and non-European observers and is compiled by the European Association of Remote Sensing Laboratories

(EARSeL) [494], with support from the ESA and the Parliamentary Assembly of the Council of Europe. Over 170 members from European agencies and institutions concerned with remote sensing are listed on a country-by-country basis and the *EARSeL Directory* is also indexed by subject fields, which range from land use to co-operation with developing countries.

The United Kingdom Remote Sensing Directory [23] updates the Department of Trade and Industry's *Remote Sensing of Earth Resources* and provides comprehensive details on approximately 220 U.K. organizations, groups and individuals engaged in remote sensing. Reports are provided on the facilities and activities of five main organizational categories, which include academic, commercial, industrial, consultancy and government bodies that act as sponsors of remote sensing activity. At the time of writing a largely similar publication is under active preparation by the British National Space Centre (BNSC) [715]. The *Directory of UK Space Capabilities* was published in 1987 and provides details of British academic, commercial and industrial institutions' space activities. A somewhat more dated publication, the *Directory of research and development activities in the United Kingdom in land survey and related fields* [10], was compiled by the U.K. National Group for Communication in Surveying and Photogrammetry in 1982. The addresses and profiles of research and development organizations, together with other useful contacts, are provided in a concise and accessible manner. *The remote sensing sourcebook* [259] contains some very useful information on British commercial organizations and companies, and also provides some detail on the wider North American and European remote sensing markets.

A good source of information on contemporary developments in remote sensing from Europe is the annual catalogue of *French remote sensing activities, services and products* [14]. Issued as part of a series of news publications by the non-profit-making Prospace [599] association, this catalogue is available in French and English versions. One section is composed of a loose-leaf folder that provides a colourful and informative guide to ground stations, systems, data collection, platform location and the equipment associated with sensors and data processing. Volume 2 describes the developments in onboard equipment by the various Prospace members and further volumes are to be published in due course.

The Remote Sensing Association of Australia (RSAA) [514] produced the 2nd edition of its 206 page *National directory of remote sensing in Australia* [20] in 1980. It deals with groups and individuals involved in Australian remote sensing programmes and is very similar in layout to the DTI's *Remote Sensing of Earth Resources.*

The second edition of a *Catalogue of European Industrial Capabilities in Remote Sensing* [4] was compiled by the Joint Research Centre (JRC) [502] in 1982. The list of over eighty companies from ten European countries that is provided by this directory is slightly dated, but nevertheless very useful. An intriguing aspect of this publication is its update capability, whereby manufacturers mentioned in the catalogue may update their entries by the use of videotex. Any change in entry will be displayed on the Commission of the European Communities' [505] videotex information service, and non-videotex users may receive print-outs of the videotex information if they contact the Commission at the above address.

One further publication of interest to those seeking data on remote sensing research is the annual *Jane's Spaceflight Directory* [18]. Although this is basically a manual of aerospace developments and products, it nonetheless provides concise, yet very thorough, background detail on a variety of international remote sensing programmes. A particularly admirable feature of this annual is an exhaustive table that gives meticulous details of remote sensing, military reconnaissance and other satellite launches between 1982 and 1985.

Several other alternative sources of directory-style information are available for remote sensing and although not entirely devoted to the subject, the very nature and scope of these guides makes them attractive alternatives to the more 'dedicated' sources. Of great relevance to remote sensing and related disciplines, such as space technology, is the 1984 publication *Earth and Astronomical Sciences Research Centres* [13]. This superb directory has the same format as the previous publication, but is far more specific, with over sixty very detailed organizational profiles that are directly concerned with remote sensing and its sub-disciplines. Another of these guides, *Engineering Research Centres* [2], is international in coverage and aims to satisfy the information needs of the engineering community concerning academic, industrial and governmental activity. More than 5,000 profiles of research, development and design laboratories from a diverse range of public, private and academic organizations are included in an alphabetical listing by country. There are more than sixty entries for aerial photography and remote sensing in the subject index. The 19th edition of a similar publication, *Industrial Research Laboratories of the United States* [19], deals exclusively with North American sources. There are a considerable number of references to remote sensing out of the 9,538 entries dealing with research and development, and the information is also available online. Another database service, FEDERAL RESEARCH IN PROGRESS, identifies and provides information on current research and projects completed within the last two years. This file contains research summaries, names of principal researchers and performing and sponsoring organizations, and project titles. It is available retrospectively from 1972 to 1982 on the now closed Smithsonian Science Information Exchange system, which contains 439,000 records.

Five volumes in the Longman Guides to World Science and Technology series have so far been published, respectively covering the Middle East, Latin America, China, Japan and Europe, and several more are under preparation. Perhaps the most pertinent to remote sensing is *European Sources of Scientific and Technical Information* [15], which provides scientific and technical information on specific topics in each country. The method of division is by subject and there are several entries on aerial photography, which includes photogrammetry, and remote sensing.

Publications originating within the U.K. include the biennially issued *Industrial Research in the United Kingdom* [17]. Although this guide features many entries of relevance to remote sensing in its 3,400 profiles, it has been largely superseded by *The United Kingdom Remote Sensing Directory*. Perhaps the most useful general source of information on remote sensing activities in the U.K. is *Current Research in Britain: Physical Sciences (CRIB)* [7], which was first published in 1986 by the British Library Document Supply Centre (BLDSC) as a successor to its *Research in British Universities, Polytechnics and Colleges (RBUPC)*. *CRIB* is a comprehensive guide to ongoing research that has an excellent coverage of academic institutions involved in remote sensing, aerial photography and photogrammetry.

3.4 Dissertations and Theses

One of the most frequently ignored sources of current information in any scientific field is the doctoral thesis or dissertation. It is possible that many researchers do not consult this type of information as they are unaware of the ease with which theses may be located and obtained. Theses or dissertations are specialized and contemporary data sources that have been prepared as an original contribution to knowledge in their particular field. Doctoral dissertations are generally concerned with precisely bounded areas of research and can provide timely communication of research results as well as alerting potential doctoral students to areas of compromise. Many dissertations will contain extended literature reviews and historical accounts that can usefully summarize the development and state-of-the-art in a particular topic.

The major publication dealing with dissertations and scientific theses is *Dissertation Abstracts International (DAI)*, which is issued in two parts, with *Dissertation Abstracts International, Section B: Physical Sciences and Engineering* [9] being the most relevant to remote sensing. *DAI* consists of more than 2,500 author-prepared abstracts each month, from over 400 North American and foreign academic institutions. The publishers, University Microfilms International (UMI) [792], also produce the *Comprehensive Dissertation Index* [5], which lists over 417,000 dissertations in 37 volumes. Its ten-year cumulation, 1973-1982, is an excellent source of information, containing more than 351,000 dissertations in a 38 volume series. These sources, together with *Masters Abstracts* (quarterly) and *American Doctoral Dissertations* (annual), are available in printed or microfiche formats.

DISSERTATION ABSTRACTS ONLINE, the computerized version of *Dissertation Abstracts International* and the *Comprehensive Dissertation Index*, cites nearly all of the American dissertations granted by accredited academic institutions since 1861. Over 800,000 items are included in the database and an average of 3,500 records are added at each monthly update. The coverage embraces all subject areas and author-prepared abstracts have been added to those records entered onto the database since 1980. As part of its Dissertation Information Service, UMI also produces short, subject-specific catalogues of doctoral dissertations taken from DISSERTATION ABSTRACTS ONLINE using a simple search strategy. These catalogues appear on an irregular basis and review a certain time span in each issue. Two of the most recent, *Research in meteorology and climatology 1975-1986* [22] and *Research in geography 1976-1986* [21], contain remote sensing as a subject sub-discipline.

Meredith and Sacks (1986) have produced an in-depth study of education in environmental remote sensing that consists of a bibliography and characterization of doctoral dissertations. Building on Meredith's 1981 paper concerning doctoral dissertations in remote sensing and photogrammetry, this study lists 356 dissertations granted by North American universities and colleges between 1965 and 1984. The results of the analysis are grouped into twelve topic areas and a summary of the award-granting institutions is provided. Peaks were noted in 1976 and 1983 when the numbers of dissertations granted rose dramatically. A similar search of DISSERTATION ABSTRACTS ONLINE, to determine numerical patterns in thesis awards, by Hyatt (1986), gave broadly similar results. For the ten-year period 1975-1985, 314 dissertations were awarded and 258 of these were in the 1980-1984 period. The number of theses granted nearly doubled and then trebled in 1976 and 1980 respectively, reflecting a growing research capability and awareness in remote sensing.

The BLDSD's publication *British Reports, Translations and Theses* [3] announces newly received theses from British universities in its monthly issue. Theses from this source have been included on the online System for Information on Grey Literature in Europe (SIGLE) [323] since 1983. An alternative and more comprehensive source is the expansively titled bi-annual *Index to Theses Accepted for Higher Degrees by the Universities of Great Britain and Ireland, and the Council for National Academic Awards* [16], published by Aslib. This publication lists the origins and accessibility of theses from all disciplines between 1950 and the present day.

The *ITC Journal* [71], which is reviewed later in this chapter, regularly publishes titles and abstracts of doctoral and MSc degrees that have been awarded to its students. A similar, but more occasional service is provided by the American Society of Photogrammetry and Remote Sensing (ASPRS) [740], which publishes lists of American dissertations granted in remote sensing on an irregular basis in the Society's journal *Photogrammetric Engineering and Remote Sensing* [99].

3.5 Journals

The range of regular and irregular serial publications that are concerned either directly or indirectly with remote sensing is very wide and a large proportion of the journals that regularly feature articles or reviews on remote sensing were originally photogrammetric in nature and outlook. The rapid expansion in both the theoretical and practical aspects of remote sensing, allied with a commensurate growth of written material generated by that research, has resulted in the dissemination of material over a broad publication spectrum, and it is still difficult to ascertain whether or not some journals, particularly foreign-language publications, have a substantive and well structured remote sensing content. A series of papers that may help alleviate this problem were published by Walker (1984), Thompson (1984) and Ayeni (1984) at the International Society of Photogrammetry and

Remote Sensing (ISPRS) Congress in Rio de Janeiro. These papers respectively present the current state of serial publications in western Europe, North America and Africa. All of the papers are informative and carry breakdowns of the anticipated journal readership, together with an idea of how much of the individual serial's content is concerned with remote sensing and related disciplines.

Probably the most important publications concerned with remote sensing are North American and European in origin; these periodicals are described later in this section and have a very high standard of production, content and organization. As the technology of remote sensing becomes more firmly established, the journal and periodical publications are being increasingly refined and currently feature the range of research, commercial, professional and, more recently, general publications that evolve around any technology. New journals are currently appearing at the rate of two to three per year and some of this expansion is constructive for a subject that is diversifying almost as quickly as it matures; however, unless specialist fields are the subject of interest, it would seem more expedient to accommodate the volume of papers and articles being generated within existing journals, either by expansion of those media, or by greater selectivity with regard to publication content.

Several journals are published in North America, and perhaps the most prestigious and influential of these is *Photogrammetric Engineering and Remote Sensing* [99], issued by the ASPRS [740]. This journal has been produced since the Society was founded in 1934 and has been issued with increasing frequency from that time: initially it was quarterly, then bi-monthly, and it reached its present status of a monthly journal in the 1970's. Originally appearing as *Photogrammetric Engineering*, it changed its name in 1975 to *Photogrammetric Engineering and Remote Sensing* in recognition of the rapid acceleration of interest in remote sensing systems and applications. *Photogrammetric Engineering and Remote Sensing* is an international journal that is distributed to over 10,000 subscribers and Society members. A wide variety of articles are featured, including engineering reports, practical papers on the applications and theory of photogrammetry and remote sensing and some very technical contributions, a large proportion of these papers are from authors outside the United States. There are interesting sections concerned with professional placement, news, forthcoming meeting announcements and interviews with important figures in photogrammetry and remote sensing.

Owing to its very widespread distribution and reputation as a leading remote sensing journal, *Photogrammetric Engineering and Remote Sensing* is regarded as an excellent advertising medium. Products and services are publicized in adverts and a state-by-state 'Professional Directory' informs the potential customer of where various companies are located. Special issues are published periodically and at the time of writing two of the most topical of these are *Photogrammetric Engineering and Remote Sensing August 1985, SPOT*, a collection of selected papers from the SPOT Applications Symposium in Scottsdale, Arizona, and *Photogrammetric Engineering and Remote Sensing September 1985, LIDQA (Landsat Image Data Quality Analysis)*, concerning Landsat 4 and 5 TM and MSS data utilization.

Remote Sensing of Environment [104] is another journal dedicated to remote sensing, which features special issues and reviews. Published on a bi-monthly basis by Elsevier, *Remote Sensing of Environment* provides up-to-date and interdisciplinary coverage of the theoretical and applied aspects of remote sensing. The high-quality research papers that are presented are augmented by summaries of previously published material, surveys, short communications concerned with developments and tutorial articles.

The *Transactions* and *Proceedings* of the Institute of Electrical and Electronics Engineers (IEEE) [764] are probably the most comprehensive publication series available that deal directly with the technical and theoretical aspects of remote sensing and related sciences. Most relevant to remote sensing are the bi-monthly *IEEE Transactions on Geoscience and Remote Sensing* [61]. This very high-quality, large-format journal is dominated by technical papers on the theory and practice of remote sensing. A short correspondence section deals with updates to, and corrections of, previously published material. Special issues, generally the published results of conferences and meetings, appear on an irregular basis. The *IEEE Transactions on Geoscience and Remote Sensing* are produced by the IEEE Geoscience and Remote Sensing Society [763], an organization existing within the framework of the IEEE, composed of IEEE members with a primary interest in geoscience and remote sensing.

Another IEEE publication that is concerned with remote sensing research is *IEEE Transactions on Pattern Analysis and Machine Intelligence* [63]; a bi-monthly journal containing selected archival papers on all aspects of pattern recognition, artificial intelligence, image processing and their respective applications. This journal features articles concerned either directly or indirectly with remote sensing on a fairly regular basis. Several more IEEE publications contain material related to remote sensing. Two highly technical periodicals, *IEEE Transactions on Microwave, Theory and Techniques* [62] and *IEEE Transactions on Aerospace and Electronic Systems* [60], are the testing grounds of the development scientist and deal with some very advanced concepts and their application over a range of fields, many of which are complementary to remote sensing. *IEEE Transactions* have similar layouts, generally with two sections, one on papers and another on correspondence; occasionally a journal might also feature short papers or letters. Special issues, generally the result of the respective society's symposia or meetings, are often combined with the normal journal. Another North American journal, *Pattern Recognition* [96], published bi-monthly for the Pattern Recognition Society [776] by Pergamon Press Inc., is also of direct interest to those involved in remote sensing. The Pattern Recognition Society was formed to serve as a focal point in what was previously a widely scattered science. The resultant publication and its various special issues feature scholarly articles on pattern recognition and digital image processing. An equivalent international publication, *Pattern Recognition Letters* [97], produced bi-monthly by Elsevier, is the official journal of the International Association for Pattern Recognition [506]. By limiting article length this periodical can usefully offer timely, yet concise and broad, coverage of the developments and literature pertaining to pattern recognition.

The minimal delay between submission and publication of articles makes this journal one of the best in its field for the precise communication of information.

The major source of learned papers for remote sensing specialists and photogrammetrists in Canada is the bi-annual *Canadian Journal of Remote Sensing* [36], published for the Canadian Remote Sensing Society [550] by the Canadian Aeronautics and Space Institute. This journal has expanded considerably since its appearance in 1974, reflecting Canada's growing interest in remote sensing applications. Articles are of a general nature, but tend to concentrate on the more practical aspects of remote sensing; papers are abstracted in French and English.

Several publications originating from Canada and North America that are not dedicated to remote sensing and its related sub-disciplines nevertheless feature quite frequent articles with a strong remote sensing or photogrammetric bias. Although it is strongly biased towards surveying and mapping, nearly half of the regular publications in the *Canadian Surveyor* [37] are on photogrammetry and remote sensing, the well-established and commendable quarterly publication of the Canadian Institute of Surveying [549]. The American Congress on Surveying and Mapping (ACSM) [745] and the ASPRS are closely linked, an alliance which is reflected in the content of papers published in their quarterly journal, *Surveying and Mapping* [119]. This journal is circulated to over 11,000 subscribers and society members, reflecting its multidisciplinary appeal. The *Journal of Imaging Technology* [76], a bi-monthly publication of the Society of Photographic Scientists and Engineers (SPSE) [788], is concerned with the technical side of photography. The papers are probably relevant only to the photogrammetrist with an interest in photographic science and its interaction with photogrammetry. Both the *American Cartographer* [29] and *Cartographica* [38], respectively publications of the ACSM and York University, Toronto, publish infrequent articles on digital mapping. It is not intended to review the wealth of cartographic periodical and serial literature available here, as remote sensing articles appear only sporadically in these publications, with the exception of occasional special editions.

The United Kingdom and Europe in general have an excellent selection of journals concerned with remote sensing, aerial photography and photogrammetry. The premier domestic publication is the *International Journal of Remote Sensing* [68], the official journal of the Remote Sensing Society (RSS) [733], which is published on a monthly basis by Taylor & Francis Ltd. The journal is characterized by consistently excellent papers and occasional review articles. It also includes a comprehensive news section that is international in scope, a calendar of forthcoming events and a summary of remote sensing work in progress. Selected short contributions and the news sections from this journal undergo a bi-monthly reprint to appear as the *Remote Sensing Society's News and Letters* [387]. This publication, distributed free to society members, is intended to be a source via which brief advance notice of research and application papers can be given. The other major British periodical concerned with remote sensing is the official journal of the Photogrammetric Society [732], the *Photogrammetric Record* [101]. Published twice-yearly, the emphasis is on theoretical and applied aspects of photogrammetry; however, despite this bias some excellent remote sensing articles are presented. The journal contains news, book reviews, membership lists and a superb section on photogrammetry around the world, consisting of brief abstracts from papers dealing with remote sensing and photogrammetry, from a wide range of international periodical and serial publications.

The Commonwealth Association of Surveying and Land Economy [720] has published the quarterly *Survey Review* [123] since 1931, its principal role being the presentation of specialized articles on photogrammetry and surveying. The political and commercial aspects of space are deliberated upon in *Space Policy* [118], a quarterly publication of Butterworth Scientific Ltd. The journal is well balanced, with discussions on aspects of space policy, scholarly contributions on issues in space, many concerning remote sensing, book reviews and a news section that includes relevant abstracts from official documents and policy statements.

More popular journals concerned with space research and carrying a good deal of remote sensing material are now becoming available in the United Kingdom. *Space* [114], published by the Shephard Press Ltd., is a controlled-circulation publication that is free to those involved in the aerospace industry at managerial level. It reflects the commercial, rather than academic, status of space research and carries some excellent review articles on remote sensing. A considerable number of journals regularly include remote sensing articles, or concentrate on remote sensing via special issues. Of particular note is *Advances in Space Research* [27], the irregular journal of the Committee on Space Research (COSPAR) [604], published by Pergamon Press.

A number of remote sensing journals are published in Europe. One of the oldest is *Bildmessung und Luftbildwesen* [34], published for the German Society of Photogrammetry and Remote Sensing [613] since 1926; this bi-monthly journal contains technical papers, abstracted in English and French, on photogrammetry and aerial surveying. There are reviews of English-language remote sensing books and a special annual edition is published in English. Carl Zeiss [611] of West Germany produces *Zeiss Information* [127] on an irregular basis; it is principally concerned with trade and highlights developments in equipment. Another 'in-house' publication, the *Journal for Photogrammetrists and Surveyors* [77] from Jenoptik Jena GmbH [614], E. Germany, again deals mainly with trade-related information; however it also features some useful articles on photogrammetry and remote sensing.

The Dutch contribute two major publications to the periodical literature of remote sensing. *Photogrammetria* [98], the official journal for the ISPRS [501], is a bi-monthly publication of Elsevier Science Publishers, Amsterdam. Markedly international in scope, it presents informative articles, reports on the society's various congresses, meetings and activities, plus news and book reviews. Papers are occasionally accepted in French or German, although the official language of the journal is English. With a circulation of over 5,000 in 145 countries the *ITC Journal* [71], published for the International Institute for Aerospace Survey and Earth Sciences (ITC) [500] in Enschede, is the largest European journal dealing with aerospace survey and applications. One of the *ITC Journal*'s main concerns is education and training in all aspects of remote sensing and particular emphasis is attached to techniques that can be applied to developing countries. The coverage is comprehensive and includes articles on remote sensing and photogrammetry, abstracts of theses and journals, book reviews, conference reports, news and an international events calendar. In addition to the *ITC Journal*, three other irregular serials are published by the ITC. *ITC Information Booklets* [223] provide 'one off' updates on developments and specific projects at the ITC. The *ITC Publications Series A (Photogrammetry)* [224] and *Series B (Photointerpretation)* [225] supply up-to-date overviews on their respective topics, and these latter two series publications are sometimes combined in one volume.

The French bi-monthly journal *Photointerprétation* [102], published by Editions Technip, Paris, emphasizes aerial photographic interpretation; papers and case studies are mainly in French, but a brief English or Spanish commentary is also provided. The format of this journal is superb and it is probably the best application-oriented practical remote sensing periodical currently available.

The Société Française de Photogrammetrie et de Télédétection [607] has produced the quarterly *Société Française de Photogrammetrie et de Télédétection. Bulletin* [111] since 1961. The journal carries scholarly articles in French, with English and German abstracts, that are of interest to those involved in remote sensing, photogrammetry, the geosciences and surveying.

The Società Italiana di Fotogrammetria e Topografia [639] publishes the *Società Italiana di Fotogrammetria e Topografia e Teledetection. Bolletino* [109] on a quarterly basis. Many aspects of remote sensing and photogrammetry are covered in Italian and occasionally English or French. The Ministero della Finanze [638] produces the semi-annual *Rivista del Catasto e dei Servizi Technici Erariari* [108], and although it is primarily of interest to cadastral surveyors and technicians, useful papers and references on photogrammetry and remote sensing are also featured. The only other Italian publication of interest is the *Bolletino di Geodesia e Scienze Affini* [35], published by the Istituto Geografico Militare [637]. The content reflects military interests in surveying and aerial reconnaissance, and contributions are abstracted in English, French, German and Spanish.

In Belgium the official journal of the Société Belge de Photogrammétrie et de Télédétection [529], the *Société Belge de Photogrammétrie, Télédétection et de Cartographie. Bulletin Trimestriel* [110], is published biennially by H. Van Olffen of Brussels. The technical papers in this publication appear in French, Flemish and sometimes English.

The *Photogrammetric Journal of Finland* [100], issued once or twice a year and published for the Finnish Society of Photogrammetry [594] at the Institute of Technology in Helsinki, is mainly concerned with the technical aspects of surveying and photogrammetry and the journal accepts contributions in English or German.

Space Markets [117], a recent publication from Interavia of Geneva, first appeared as a supplement to their annual *Interavia Aerospace Review*. Now a journal in its own right, it aims to publicize the commercial potential in space research and development. In common with the English publications *Space* and

Space Policy it features a relaxed editorial style that readily encourages assimilation of informative articles on a wide range of space ventures.

In Poland articles and scholarly contributions to the remote sensing literature appear in the country's geodetical journals. *Geodezja i Kartografia* [56] is the quarterly organ of the Polish Academy of Science [677], and occasional remote sensing articles from Academy members are printed in it, although the journal's emphasis is on geodesy. Academy members and Polish university researchers infrequently submit remote sensing papers to other journals, which again usually have a strong bias towards geodesy.

Cartographia [619] of Hungary distribute the bi-monthly Hungarian-language journal *Geodézia és Kartográfia* [53]. Remote sensing articles are published occasionally, although the emphasis is on geographical cartography and surveying. Papers are abstracted in English or Russian and the contents pages are given in English, Russian and German. Foreign technical journals and books are reviewed and there are regular features on trade and association news.

The main remote sensing journal in the U.S.S.R. is *Issledovanie Zemli iz Kosmosa* [70], a bi-monthly publication of the Akademiia Nauk S.S.S.R. [805] in Moscow. A complete cover-to-cover translation of this journal is issued in English as the *Soviet Journal of Remote Sensing* [113] by Harwood Academic Publishers, New York. It carries technical papers on remote sensing and provides brief abstracts of forthcoming articles. Two other Russian journal issues concerned largely with geodesy, but featuring scholarly contributions on aerial photography and photogrammetry, are the bi-monthly *Geodeziia i Aerofotos'emka* [54] and an irregular publication *Itogi Nauki i Tekhniki: Seria Geodeziia i Aeros'emka* [72]. These journals are respectively produced by the Srednego Spetsial'nogo Obrazovaniia S.S.S.R., Moscow [807], and VINITI, Moscow [811].

One of the most recently launched remote sensing journals is the quarterly *Geocarto* [52], which is published in English by the Geocarto International Centre [617] in Hong Kong. This new journal, which first appeared in 1986, is international in scope and multidisciplinary in coverage, encompassing the diverse fields of remote sensing, geoscience and cartography. The excellent large-format presentation includes articles and reviews on research developments and applications. News and evaluatory notes as well as critiques on publications and products are all well documented.

The Remote Sensing Technology Centre of Japan (RESTEC) [641] distributes the *RESTEC Journal* [105] on an irregular basis. This house organ announces news of forthcoming symposia and contains well balanced scholarly contributions on remote sensing. Israel's Department of Survey [634] produces irregular *Photogrammetric Papers* [69] in Hebrew with English abstracts. The *Australian Journal of Geodesy, Photogrammetry and Surveying* [33] has been published semi-annually since 1968 by the University of New South Wales School of Surveying [523], Kensington, and although biased towards geodesy, it accommodates photogrammetric reports, many of which are inclined towards surveying. The Mexican Society of Photogrammetry, Photointerpretation and

Geodesy [655] publishes the quarterly *Fotogrametría, Fotointerpretación y Geodesia* [49], which is received by over 1,000 society members and subscribers.

The only South American journal that is concerned primarily with remote sensing is the *Centro Interamericano de Fotointerpretación Revista* [39], published since 1976 on a quarterly basis by the Centro Interamericano de Fotointerpretación (CIAF) [584]. Despite the journal's name all aspects of remote sensing are catered for and most of the published papers are from in-house sources. Articles and reviews pertaining to remote sensing can be found on an irregular basis in the bi-annual edition of *Geografia* [57], a largely geographic periodical that is published by the Instituto de Geociências e Ciências [533] of Brazil.

The status of periodical and serial publications in Africa leaves much to be desired. Ayeni (1984) made a strong case for an African photogrammetric journal in his assessment and preview of remote sensing and photogrammetry periodicals. However, one publication originating from South Africa was not considered: the *South African Journal of Photogrammetry, Remote Sensing and Cartography* [112] is issued twice a year in English and Afrikaans by the South African Society for Photogrammetry, Remote Sensing and Cartography [684]. This journal is nicely presented and reflects the high standard of research in South Africa. The cartographic content is minimal with papers and reviews being inclined towards photogrammetry and remote sensing. The only journal outside South Africa that deals exclusively with photogrammetry and remote sensing is the annual *Nigerian Journal of Photogrammetry and Remote Sensing* [90], first published in 1984 by the Nigerian Society of Photogrammetry and Remote Sensing [666]. Other serial publications that are devoted to cartography and surveying, but occasionally include articles pertaining to photogrammetry, are the *Nigerian Geographical Journal* [89] and *Nigerian Engineer* [88].

3.6 Textbooks

Some of the problems that arise with remote sensing literature, including textbooks, have already been discussed under the general characteristics and classification sections in this chapter. Remote sensing offers a very diverse book literature, although many of the texts mentioned in Part II are published conference proceedings, which are not books as such. A great many of the textbooks for remote sensing are application oriented, focussing on a wide range of practical themes, ranging from agriculture and the atmosphere to water resources and land use planning. As a discipline, remote sensing requires a great deal of background knowledge, and most of these books attempt to incorporate that knowledge as an introduction to the subject. The author does not intend to criticize this approach by other writers, but to provide a general synopsis of remote sensing when it is not the sole focus of a text can be very difficult. In some cases this type of approach tends to the almost superficial, especially as there are several excellent general introductory texts, and an equal number of more specific ones, that cover this area most adequately.

Over 100 textbooks and conference proceedings are reviewed in Part II; necessarily they cover a wide range of subjects, which are indicated below, together with their respective entry numbers. Although the author is very familiar with remote sensing literature and practices, he by no means professes to be an expert on all (or any!) of these topics. Each entry in Part II is annotated, and if the annotation contains a suggestion of relevance to potential readers' subject area(s), they are strongly advised to consult the complete work or works.

Aerial Photography [130-142]
Agriculture [143-149]
Atmosphere [150-158]
Civil Engineering and Planning [159-160]
Digital Image Processing [161-168]
Ecology [169-170]
Forestry [171]
Geography [172-174]
Geology [175-182]
Ice and Snow [183]
Introductions to Remote Sensing [184-189]
Land Use [190-194]
Microwave [195-199]
Oceans [200-207]
Photogrammetry [208-210]
Technical Remote Sensing [211-221]
Training [222-229]
Water Resources [230-232]

A new bibliographic guide to books, the *Index to Scientific Book Contents* [288], was first published in 1985 by the Institute for Scientific Information (ISI). The *Index to Scientific Book Contents* was devised in order to help researchers and librarians identify the chapter contents of multi-authored books. These publications are as common in remote sensing as other sciences and often the book's content cannot be assessed from the brief cover title that is provided. The 1985 annual provides access to nearly 31,000 chapters from over 2,000 multi-authored books; the 1986 and subsequent editions will be published in three quarterly issues and an annual cumulation that will incorporate the fourth quarter. The *Index to Scientific Book Contents* format catalogues books and chapters by contents of books and author/editor, permuterm subject and corporate indexes. The 'Contents of books' is the central index to which all others refer, and for each book it features an identifying number and bibliographic details, including each chapter author.

The *British National Bibliography* [257] is the major British source for information on newly published books. It has been available since 1950 and is currently issued weekly by the British Library Bibliographic Services Division (BLBSD) [712], with quarterly and annual cumulations. Indexing is by author and subject with Dewey Decimal Classification order. The online database that corresponds to the *British National Bibliography* is UKMARC, which currently holds nearly 930,000 citations and can be searched on the British host BLAISE. The same system may also be used to search LCMARC, the online equivalent of the *National Union Catalogue* [269], which stores book literature catalogued by the American Library of Congress (LOC) [770]. American and all other books published in English are comprehensively covered by the *Cumulative Book Index* [261], a monthly publication of H. W. Wilson, with quarterly and annual cumulations. The *Cumulative Book Index* has been issued since 1898 and is available on microfiche from University Microfilms International (UMI) [792], who also offer a reprint service for out-of-print works. The machine-readable version of the *Cumulative Book Index* carries material from 1982 to the present and contains over 200,000 citations, making it one of the most current bibliographic records available. The database is updated every two weeks and new books tend to appear here before anywhere else. Another useful publication is *Books in Print* [255], produced since 1947 by R. R. Bowker. An annual list of titles in print is provided in six volumes; a quarterly microfiche is available from 1982, and the information may also be accessed online.

3.7 Reports

The reports literature of remote sensing is as wide and varied as that of any other subject, with the various academic, commercial and governmental organizations involved in remote sensing or related fields publishing a diverse selection of material. Much of their output is in the form of standard annual reports, which are tabulated in the Journals section above. The types of reports these groups generate include technical, committee and governmental reports, together with more specific research development reports. Many report issues chronicle the methodology and results of research that is initiated periodically and this is reflected by their somewhat irregular publication frequency. These irregularities in production pose a problem for libraries or bodies that wish to keep informed of the reports literature. Reports are notoriously difficult to classify and catalogue as very few appear in a structured series and this problem is exacerbated by the variety of publication formats, generally paper or microfiche, which in themselves necessitate different storage criteria.

Some established report series are recognized for remote sensing, notably the publications *NASA Technical Reports* [238], *NASA Technical Memorandums* [236], *NASA Technical Notes* [239] and *NASA Technical Briefs* [239]. The arrangement of this kind of technical publication is often repetitive, being reproduced for succeeding issues, which can aid the reader who is familiar with a report's format in the ready assimilation of its content. Several journals, such as SUPARCO's *Space Horizons* [116], regularly feature report summaries on specific projects that have been undertaken in their parent organization.

The nature of reports, which often include summaries, detailed analyses, conclusions and recommendations, makes them particularly pertinent reference sources. Many reports do however suffer from a reduced availability or restricted circulation which usually stems from a reluctance on the side of the aerospace industry and remote sensing industry to publicize its work, a great deal of which is defence oriented. With regard to remote sensing, the greatest cause of restricted circulation is probably the potential commercial value of a report, as developmental and market research reports derived from private-sector research represent considerable investments by the organizations that initiated them. These reports may be of little value two to three years after publication and they can then be released; however, such time delays will not encourage the systematic collection of this material by libraries.

The best way to keep abreast of the available reports literature is to refer to the major report announcement serials. The principal publication that contains report abstracts of remote sensing interest is *Government Reports Announcements and Index* [235], a bi-monthly publication of the National Technical Information Service (NTIS) [799]. This journal may be used to order reports from a variety of organizations through NTIS, and it includes details such as personal author, corporate author, keyword, contract or grant numbers, title and NTIS order number, together with a brief abstract of the report's content. An example of the complete format that entries appear in is provided below.

631,349
N86-20934/3/GAR PC **A02**/MF **A01**
National Aeronautics and Space Administration, Moffett Field, CA. Ames Research Center.
Simulation of LANDSAT Multispectral Scanner Spatial Resolution with Airborne Scanner Data.
C. A. Hlavka. Jan 86, 10p NAS 1.15:86832, A-85400, NASA-TM-86832

A technique for simulation of low spatial resolution satellite imagery by using high resolution scanner data is described. The scanner data is convolved with the approximate point spread function of the low resolution data and then resampled to emulate low resolution imagery. The technique was successfully applied to Daedalus airborne scanner data to simulate a portion of a LANDSAT multispectra scanner scene.

Government Reports Announcements and Index has a broad coverage of the scientific and technical, as well as business and economic report literature generated by U.S. government agencies and contractors, together with some external foreign contributors. The *Announcements* are arranged in twenty-two subject categories, which are in turn divided into 178 sub-categories. An online version, the NTIS [318] database, which can be searched through a variety of hosts, is fully described in Section 4.5.

Further information on reports concerned with remote sensing may be obtained from *Scientific and Technical Aerospace Reports* [240]. Available from the U.S. Superintendent of Documents [791], *Scientific and Technical Aerospace Reports* is a bi-monthly announcement of NASA, NASA contractor and NASA grantee reports, together with the reports generated by other U.S. government agencies and institutions. Reports, patents, translations and theses from a selection of both

U.S. and foreign academic, commercial and governmental organizations are also covered and references to remote sensing may be identified in several of the seventy-four subject categories, which are divisions of the ten major subject entries. It should be noted that repetition can occur with announcement journals and many of the reports cited in *Scientific and Technical Aerospace Reports* can also be found in *Government Reports Announcements and Index. Scientific and Technical Aerospace Reports* is one of the two abstracting journals used in the compilation of the AEROSPACE [307] online database, which is composed of report literature and other government documents.

The major announcement journal in the U.K. is *British Reports, Translations and Theses* [3], published on a monthly basis by the British Library Document Supply Centre (BLDSC) [713]. In contrast to the above-mentioned American publications, the entries in this serial are purely bibliographic in nature, with no abstract being provided. The content of *British Reports, Translations and Theses* reflects the report collection at the BLDSC, which also houses a great many of the reports from NASA and various other American institutions. NASA reports are also submitted to specified university and public library facilities in the United States, while in Europe, NASA reports are archived by the Information Retrieval Service (IRS) [498] of the ESA, the interests of NTIS being represented by Microinfo Ltd. [728] of Hampshire.

Although the reports literature displays notable weaknesses in some areas, it is nevertheless one of the principal avenues via which groups or individuals with an interest in ongoing developments in narrow specialist fields may remain informed. In some areas the report may stand alone as an authoritative work; however, it is often followed by the more inwardly digestible and conventional forms of communication, such as learned papers and articles.

3.8 Conference Papers

Conference and meeting papers represent excellent current and retrospective sources of information; however, gaining access to these media can be difficult at times. The major remote sensing conferences, symposia, workshops and meetings that are held on a regular basis are discussed in Section 4.7, together with details of the announcement services that provide prior warning of these events. The theoretical and application-oriented papers presented at remote sensing conferences are usually subject to excellent bibliographic control and the proceedings generally appear in bound volumes subsequent to the meeting, although in the case of some conferences, proceedings are issued at registration. In some cases the results of a society's special meetings or symposia may occasionally be published in the official journal of that society, and special issues such as these have been discussed earlier in this chapter. A lot of the practical and technical themes from the various conferences are iterative, and it is relatively easy to gain a broad outline of a specific topic at an introductory level by careful selection of material. It should however be stressed that many of the papers presented are of the highest technical order and often appear in print far more rapidly than submissions to various specialist journals.

The announcement series for conference literature generally impart information in two ways: either a brief summary of a conference's scope and coverage, with no listing of individual papers, or individually listed papers which occasionally have brief abstracts. An example of the latter method is the *Conference Papers Index* [241], which is a monthly publication from Cambridge Scientific Abstracts. The information sources for this serial include abstracting journals and final conference programmes, which can sometimes be disappointing, as many papers are submitted for printing too late, and only their abstract appears in a proceedings. The online version of *Conference Papers Index* carries records back to 1973 and covers over 110,000 scientific papers that have been presented at more than 1,000 conferences each year. The database, which may be searched through ESA-IRS and DIALOG, lists many papers prior to publication and provides useful advice on their location. The *Index to Scientific and Technical Proceedings* [245] is published every month, with an annual cumulation, by the Institute for Scientific Information [765]. It is similar to the previous *Conference Papers Index* as it lists individual papers. The online version, ISTP&B, contains over 1 million citations that provide a multidisciplinary index to most of the relevant proceedings literature. Both of these sources are searchable by a variety of types, including conference subject, location, date, sponsor, author and title indexes.

A British publication, *Index of Conference Proceedings Received* [244], takes a distinctly different line to the latter sources, as it lists conference proceedings volumes received by the BLDSC [713], instead of individual papers. This serial, which is issued on a monthly basis by the BLDSC, with five, ten and eighteen-year cumulations, features an alphabetical arrangement, subject keywords and a brief summary of a conference's topic. Its corresponding database, the CONFERENCE PROCEEDINGS INDEX, is based on the 190,000 conference proceedings volumes held by the BLDSC, including the *British Lending Library Conference Index 1964-1973*, *Conference Index: Five year cumulation 1974-1978* and the recent supplement *Index of Conference Proceedings Received.*

One of the most comprehensive guides to conference papers is the *Ei Engineering Conference Index* [242], which provides rapid and easy access to over 1,600 worldwide conferences and meetings. The first edition of this publication consists of six volumes and an accompanying two-volume cumulative index. Remote sensing is represented in several volumes, which are themselves divided into the various sub-disciplines of engineering. The *Ei Engineering Conference Index* is based on citations included in the online database EI ENGINEERING MEETINGS from July 1983 to July 1984. The database was initiated to keep pace with the growing literature generated by engineering and technical conferences and every paper is included from about 2,000 worldwide meetings each year. A review record of each conference provides a summary of its scope and also appears in the companion file COMPENDEX [310], which facilitates cross-file searching between the databases.

3.9 Trade Literature

The 'trade literature' is a profuse and diverse collection of those publications that are concerned wholly, or in part, with publicity. These media are notoriously difficult to catalogue and classify as they consist of journals and catalogues at the top end of the scale and range down to newsletters, pamphlets and publicity sheets at the lower end. If the vagaries and irregularities of publication are also considered, the task of keeping abreast of this literature is seemingly impossible. Libraries and information centres will often neither stock, nor order, such publications, as their commercial nature makes them very short-lived in terms of literature usage. In many cases the only way to acquire regular updates on this type of information is to have previously ordered or enquired about a product or service available from the issuing organization. Indeed those groups that have taken advantage of an organization's facilities or products may find themselves in receipt of liberal amounts of free publicity material.

In remote sensing there are very few 'trade only' journals and those that do exist are mainly concerned with the hardware aspects of photogrammetry. The *Jena Review* [74] and *Journal for Photogrammetrists and Surveyors* [77] are German journals that provide details of advances in photogrammetry, while mainly concerning themselves with the new and established optical and photogrammetric instruments available from Carl Zeiss [611]. The same company also produces *Sensing the Earth* [252], a useful and informative catalogue that consists of various data sheets on the hardware products and instruments made by the issuing organization.

In the United States, the quarterly journal-cum-newspaper, *Photogrammetric Coyote* [379], provides an entertaining and informative review of the services and facilities of E. Coyote Enterprises Inc. [758]. A broadly similar French bulletin, *News from Prospace* [375], features detailed information on products used in French space activities and is particularly concerned with the Ariane and SPOT programmes. Prospace [599] also publish a variety of catalogues that have a strong trade bias and these have been mentioned earlier, with regard to research in remote sensing.

A recent and potentially very useful innovation for the aerospace engineer is the publication of six product data books by Technical Indexes Ltd. [738]. These volumes are specifically geared towards aiding the user in locating supplies of over 250,000 engineering products from 21,000 U.K. companies and their respective overseas agents. The indexes are keys to accompanying microfiles, or 'prime sources files', that provide expansive details from catalogues in their particular subject field. The two *Indexes* of most relevance to those interested in remote sensing equipment are probably the *Electronic Engineering Index (EEI)* [251] and the *Construction and Civil Engineering Index (CĊEI)* [250]. A fairly similar service is provided by the Videolog facility on the IEEE FINDING YOUR WAY [314] online database. This feature allows the perusal of over 650,000 electrical and electronic components from over 700 manufacturers, as well as news of the latest developments in trade fields.

The impact that technology is having in this field is quite marked and well illustrated by two databases, SATELDATA [321] and SPACECOMPS [326], that are of interest to organizations involved in trade and industry. SATELDATA provides information on satellite hardware, listing technical parameters of past, present and future space systems, while SPACECOMPS deals with space componentry, which ranges from solders and fuses to instrumentation and integrated circuits.

3.10 Patents

Patents are probably some of the most widely overlooked sources of practical, concise and precise technical information currently available. Very few people engaged in remote sensing research will ever consult the vast patent literature that is available, unless they are fundamentally concerned with aerospace engineering and design. By definition patents are detailed legal and technical documents; in addition, and with regard to science and technology, they are new ideas and feature innovative steps that should usually have an industrial or commercial end use.

Patents may be the first, and often the only, published account of an invention, as their content is infrequently translated into a more easily understood form. The value of this type of literature is realized upon the appreciation that patent structure follows a specific and repetitive format, which can lead to rapid assimilation of content once the basic make-up has been determined. A great deal of concise and detailed information about the history and current state-of-the-art of the field into which a patent enters is often included as part of the patent specification and the drawings and account of the problems the invention hopes to solve combine very well with this narrative. For remote sensing and related disciplines, patents are a useful information source that may review and preview developments in the hardware aspects of aerospace technology. By association of these ideas, a patent can indicate future advances not only in instrumentation, but in the application-oriented aspects of the science.

It is not difficult to search the patent literature, as most of it is available through a range of online database systems. Hard-copy equivalents of these are published and these may be useful for historic searches; however, it is probable that most patents relevant to remote sensing are quite recent and therefore more accessible through computer-readable databases. One of the major files is INPADOC, which has been available on a limited basis since 1968, but more comprehensively from 1983. It may be searched through INKA and PERGAMON INFOLINE, and over 1.5 million patents from fifty-one patent authorities are included on the file, which is very useful for checking a subject area's patent coverage on an international scale. The WORLD PATENT INDEX, although available from 1963, only includes abstracts of filed material from 1981 onwards, and nearly 4 million patents from over thirty countries, in all subject disciplines, are stored in this file. A relatively new database, SPACE PATENTS [327], is concerned with the various aspects of space technology, such as communications, command and tracking. Initiated in 1985 and maintained as a current file, it contains over 10,000 records, copies of which are available from the respective cited country's National Patent Office.

Two major sources of information on U.S. patents are *Patent Abstracts* and *Patsearch*, which respectively list mechanical and electrical patents since 1963 and patents covering all areas of technology served by the U.S. Patent and Trademark Office [800]. Many of the equivalent hard-copy publications of these databases are listed by Fisher (1986). Several journals that include remote sensing articles occasionally publish a list of patents. The *ESA Journal* [48] regularly lists patents applied for by ESA, and *Applied Optics* [31] features patent abstracts, which are sometimes accompanied by descriptive line drawings and diagrams.

It is probable that few individuals will consult the patent literature, even in consideration of the relatively easy access to the information via online sources. Larger organizations may have the resources to employ 'patent agents' to do their searching, and most countries have centralized patent offices which have substantial collections of foreign and national patents available for consultation by interested parties.

References

Ayeni, O. O. (1984). Present state of periodicals and other series publications on photogrammetry and remote sensing in Africa with an outlook into the next decade. *International Archives of Photogrammetry and Remote Sensing* XXV (A6): 53-64

Fisher, J. W. (1986). 'Patent specifications.' In L. J. Anthony. *Information sources in Engineering*. London: Butterworths.

Hyatt, E. C. (1986). *An analysis of dissertations in remote sensing and related disciplines.* Unpublished internal database search.

Meredith, R. W. (1981). Doctoral dissertations pertaining to remote sensing and photogrammetry: a selected bibliography. *Photogrammetric Engineering and Remote Sensing* 47 (5): 617-629

Meredith, R. W. and Sacks, A. B. (1986). Education in environmental remote sensing: a bibliography and characterization of doctoral dissertations. *Photogrammetric Engineering and Remote Sensing* 52 (3): 349-365

Thompson, M. (1984). Periodicals and other series publications on photogrammetry and remote sensing in North America. *International Archives of Photogrammetry and Remote Sensing* XXV (A6): 300-306

Walker, A. S. (1984). The present state of periodicals and other series publications on photogrammetry and remote sensing in the countries of western Europe, with an outlook into the next decade. *International Archives of Photogrammetry and Remote Sensing* XXV (A6): 307-317

4 Keeping Up to Date in Remote Sensing

4.1 Literature Guides

Remote sensing is not a subject that is widely endowed with guides to its literature and until recently those seeking information on the subject have been obliged to look at sources that deal with wider themes, such as aeronautics and engineering. Literature guides in a discipline that has expanded as much as remote sensing has in recent years must be something of a compromise, because a proliferation in the published results of research, allied with the widespread use of machine-readable databases, means that compiling a comprehensive bibliography of remote sensing would be a massive and somewhat pointless task.

A guide that identifies the key sources of literature information must by necessity have some concessions to the volumes of material available and it is for this reason that bibliographies and similar compilations are so useful. Most of the bibliographic sources described in this section contain no annotations and are simply author, title and source listings. For this reason they make excellent current awareness media as their inherent simplicity means they can appear in print soon after the original full-length publication.

The most recent publication to provide a guide to a range of remote sensing materials is *The remote sensing sourcebook: a guide to remote sensing products, services, facilities, publications and other materials* [259]. Published in 1986 by Kogan Page and McCarta Ltd., the book is in three parts, the first of which provides an introduction to remote sensing and an overview of its organizational structure within the U.K. Part 2 emphasizes sources of imagery and the products and services, which includes publications, that are available from the remote sensing community, while the final section details educational and training facilities and organizations. The guide's content is directed toward the U.K. remote sensing community, although a limited attempt has been made to embrace the North American, and to some extent the European, markets. The information content is high; however, ready assimilation of the facts is not encouraged by the poor layout and a somewhat scanty index. Products and their prices are exhaustively listed, and although this is useful, regular supplements would be required in order to keep this sort of information up to date. Some of the major publications on remote sensing are reviewed, although in the majority of cases concerning textbooks and monographs no annotations are provided. The obvious technical literature of remote sensing is very well documented, but current awareness techniques, and online services in particular, have little or no mention. In conclusion it can be stated that this is a good reference aid which can serve as a useful sourcebook; however, it may be advisable to check the sources quoted as some are out of date.

Although more than ten years old, the only other pertinent reference book that addresses remote sensing literature is *Remote sensing of earth resources: a guide to information sources* [258], published in 1976 by the Gale Research Company. This book's layout is excellent, featuring a clearly ordered numerical listing of information sources. The annotated entries cover general literature, proceedings, manuals and guides, catalogues, maps, bibliographies, journals, workshops and training courses. There are 378 cross-referenced items which are indexed by author, title, subject, NTIS Access number, and series, while a useful list of acronyms and abbreviations is also provided. In any book of this type the annotations can only give an indication of content, and this is reflected by the reviews in this volume, which are general, but nevertheless very useful. One of the features of the guide is its lack of narrative, which focusses the reader's attention on the brief entries, although a little more explanation could have saved time in searching the references. It is perhaps only in the sections on education and training where the guide's age really shows, for many of the entries for this section are out of date. Generally though, the information contained in the book is remarkably relevant, with the bibliographic section in particular providing useful and still pertinent information, together with other data that can be used in a retrospective or historical manner. Of the two literature guides the layout of the latter volume facilitates ready access to information, whilst the former's information content is very good. In conclusion it would seem realistic that a combination of the styles is most appropriate as, in view of the advancements in remote sensing, to attempt the production of an up-to-date version of *Remote sensing of earth resources: a guide to information sources* would be a massive undertaking.

A recently produced guide to information sources, but not a literature guide as such, has been published by the USGS. The *Guide to obtaining USGS information (USGS Circular 900)* [443] describes the various information groups that operate within the USGS, and the products and services available from them. Its inclusion here is justified by the fact that one of the information centres is the USGS Library, which has over 1 million books and 300,000 maps in publicly accessible libraries situated in several States.

4.2 Bibliographies

There are a number of useful bibliographic publications that deal with remote sensing, the longest running of which is *Earth Resources: A Continuing Bibliography with Indexes* [265]. This publication is a continuation of *Remote sensing of earth resources: a literature survey with indexes* [270 (a)], published by NASA in 1970. The early survey lists 3,684 reports, papers and articles on all aspects of satellite remote sensing, and the catalogue of literature originates from documents entered onto the NASA scientific and technical database between 1962 and 1970. An extension of the 1970 publication, *Remote sensing of earth resources: a literature survey with indexes 1970-1973 supplement* [270 (b)], was published by NASA in 1975. This is essentially a continuation of the previous publication, and presents nearly 5,000 selected database entries. This guide was issued in two volumes, which are divided into several sections, the first containing citations with abstracts and the second featuring subject, author, source, contract number and report number indexes. The modern equivalent, *Earth Resources: A Continuing Bibliography with Indexes*, has been published quarterly by NASA since 1974, and in common with the previous bibliographies it features annotated citations from the NASA Science and Technology Information System. The reports and journal articles it carries are unclassified and many are the same as those presented in *International Aerospace Abstracts* [289] and *Scientific and Technical Aerospace Reports* [240]. The five indexes which are included cover subject, author, corporate source, contract number and accession number.

A useful bibliography that has been published annually since 1875 is the July issue of the monthly German journal *Zeitschrift für Vermessungswesen* [129]. Over 200 serial publications from a range of countries are scanned to produce the bibliography, which contains over 2,500 entries each year. Although the entries are largely concerned with surveying and mapping, there are useful references to photogrammetry and to a lesser extent remote sensing.

One of the largest bibliographies on remote sensing compiled by an individual is P. F. Krumpe's *The World Remote Sensing Bibliographic Index* [268]. Published in 1976 by the Remote Sensing Unit of Tensor Industries Inc., this index contains over 4,000 references for the period 1970-1976. The publication is geographically indexed by country and is slightly biased towards natural and agricultural resources throughout the world. The citations, which have no abstracts, are divided into fourteen major disciplines, with entries coming from over 850 sources in more than 150 geographic regions. The references have to be referred to the 'consecutive source code index', as they are incomplete by themselves. Guidelines are established to aid the reader in obtaining the cited documents and there are institute and publication indexes. Although this guide is dated, it is still a useful retrospective bibliography and many of the citations are still of direct interest to the researcher in remote sensing.

A dated, but by no means obsolete, guide to photogrammetric articles is *The ITC International Bibliography of Photogrammetry* [274]. Published in 1976 by the International Institute for Aerospace Survey and Earth Sciences (ITC) [500] and containing over 1,000 annotated entries, it is of particular use to those with a historical interest in photogrammetry. The compilation provides descriptions of the techniques, theory and applications of aerial surveys, photogrammetry and related fields.

There are a number of other bibliographies of use in remote sensing: *RESENA Bibliographies* [271], *Vance Bibliographies* [273], and the *Bibliography of Remote Sensing Publications* [254] from Kansas University [786] are examples. The RESENA group's publications are generally on specific topics in remote sensing, while the Kansas University series reflect publications by the staff of the Remote Sensing Laboratory. It should be noted that most of the large remote sensing organizations prepare listings of in-house publications. They are all usually subject-specific and the references cited at the end of these documents will generally provide an appreciable coverage of the topic. Many of these are detailed in Parts II and III of this book as they are neither large nor regular enough to warrant inclusion here.

A very important source, particularly for earth scientists, is the *Bibliography and Index of Geology* [253]. This consolidated list of papers has been published under several different titles and by different organizations since 1933 and is now issued on a monthly basis by the American Geological Institute [746]. The guide has a list of serials cited in each issue, together with subject and author indexes, as well as a section on fields of interest. Citations concerned with remote sensing are mainly found in the Geophysics subject division, but others may be discovered by looking under remote sensing in the 'field of interest' section. An annual cumulation is usually published in January and all the bibliographic entries are available through the GEOREF database [313]. This online service stores items dating from 1965 and currently holds nearly 1 million bibliographic citations; over 5,000 additions are made to this figure with each monthly update. Photocopies of documents or maps that are cited in the *Bibliography and Index of Geology* are available from the GeoRef Document Delivery Service (GDDS) [760]. Orders are accepted by mail, telex, telephone or online via ORBDOC or DIALORDER and express delivery services are available for an extra fee.

When searching for bibliographic information on a specific topic, dissertations, theses and journal review articles are useful, but largely overlooked, sources. Details of how to locate and obtain dissertation and thesis material, together with a list of reviewing journals and other media can be found elsewhere in this book.

Publications by the British government are noted in the *Daily List of Government Publications from HMSO* [264] and the *HMSO Monthly Catalogue* [266], which features an annual cumulation. Official government documents not published by HMSO are recorded in the *Catalogue of Official British Publications not Published by HMSO* [260], published since 1980 on a bi-monthly basis by Chadwyck-Healey Ltd. American govermental publications not usually available through the National Technical Information Service (NTIS) are listed in the *Monthly Catalog of United States Government Publications* [290], distributed by the U.S. Government Printing Office. The *Catalog* has been available online since

1976 as the GPO MONTHLY CATALOG and may be searched on BRS or DIALOG.

Periodical Bibliographies

A comprehensive coverage of all periodical publications is provided by *Ulrich's International Periodicals Directory* [272], an annual publication, with quarterly updates, from the R. R. Bowker Company. The same publishers issue the annual *Irregular Serials and Annuals: an International Directory* [267] and *The Bowker International Serials Database Update* [256], which contains all the data necessary to keep current between editions of the former two titles. Together, these publications provide bibliographic details on over 100,000 journals in 557 subject areas. Care should be taken when searching for remote sensing periodicals as they are found under a number of headings, which include Aeronautics and Space Flight, Civil Engineering, Earth Sciences, Geography, Conservation, Environmental Studies, Instruments and Electronic and Electrical Engineering. The directories are compiled from the ULRICHS PERIODICALS database, which may be searched through BRS and DIALOG, to provide coverage of over 135,000 serial publications and 65,000 periodical and serial publishers from nearly 200 countries. Another excellent source that has a comprehensive coverage of the world's periodical literature is the weekly edition of the *British National Bibliography* [257], which contains the titles of the first issues of new journals. *Current Serials Received* [263], published by the BLDSC [713], lists over 70,000 serial titles received by the British Library, and the titles of cover-to-cover translations of foreign-language periodicals are also listed.

Other Bibliographies

Several other sources outside of the normal trade press publish consolidated bibliographies that detail journals of particular interest to remote sensing and related fields. These sources include the lists of periodicals that are scanned for online database or abstracting and bibliographic sources. An example is the AIAA Technical Information Service, which publishes an annual list of over 1,600 periodicals scannned for *International Aerospace Abstracts* [289]. Libraries often have lists of their holdings: the U.K. National Remote Sensing Centre (NRSC) [703], for example, produces a list of *Current Information Received* [262] that has been compiled by Linda Newbold, the librarian. This document covers all archived and current material at the NRSC, including serial publications. It is probable that libraries of other remote sensing organizations may be prepared to provide similar compilations on materials received if they are approached.

4.3 Abstracts and Indexes

Unlike bibliographies and indexes, abstracts generally appear in the literature some time after the original full-length paper or article has been published. The very nature of an abstract requires a careful review of the material from which it is taken, which by necessity takes far longer than a brief author and source bibliographic citation. The quality of abstracting periodicals that contain remote sensing information is high, reflecting the expert skills that are being utilized in their compilation. By definition abstracts should give a considered and precise evaluation or critique of the original work. Contemporary abstracts tend to be indicative and informative, giving a brief précis of the original work, while at the same time trying to provide sufficient information to aid the reader in the decision of whether or not to consult the full-text document. Critical abstracting requires greater skill or time and is not often used with remote sensing information.

Several journal and newsletter publications include abstracting and indexing sections, and although these are not particularly large, they nevertheless present useful data to a wide readership. Of particular note is the *Photogrammetric Record* [101], which abstracts articles from an international range of periodical and conference literature. Both the *Israel Remote Sensing Information Bulletin* [363] and *Space Research and Technical Information Bulletin* [391] contain useful bibliographic sections, together with indexes, abstracts and information that has been obtained from online sources.

The most comprehensive abstracting service for remote sensing is the bi-monthly *Geographical Abstracts G: Remote Sensing, Photogrammetry and Cartography* [287]. This publication has been produced since 1974 under the name *Geo Abstracts G: Remote Sensing, Photogrammetry and Cartography* by Geo Abstracts Ltd. [724]; the new title was first used in 1986. The subject coverage of the journal is very wide and embraces all remote-sensing-related disciplines, with over 450 abstracts being included in each edition and subsequently indexed in the *Geographical Abstracts. Annual Index* [286].

The topics shown below are abstracted in each edition; being taken from the available literature, together with free copies of books, serials and conference proceedings.

Annotations generally appear in the following form:

86G/0897 Resource management trends in India using remote sensing techniques. L. R. A. NARAYAN & SEELAN SK, in: *Integration of remote sensed data in GIS for processing of global resource information. CERMA proc. 1985, Washington, DC*, (SES Inc., Springfield, VA), 1985, pp 5.10-5.16.

With the success achieved from various experiments conducted around the country, the potential applications and the future prospects of remote sensing vis-a-vis resource management on a national level has been well understood. In parallel, there have been considerable developments in sensor, satellite and launching technology. This is expected to culminate in the Indian Remote Sensing (IRS) series of satellites, starting from 1986. Several governmental departments are involved with the evolution of the system and the primary responsibility rests with the user departments. -Authors

Expansion of the Geo Abstracts Ltd service means that the contents of *Geographical Abstracts G: Remote Sensing, Photogrammetry and Cartography* are indexed on the GEOBASE [311] database, which went online in 1987. Geo

Abstracts Ltd also publish and distribute titles for the International Cartographic Association (ICA) [507], many of which are relevant to remote sensing.

Remote sensing - general
Photogrammetry and orthophotography - general
Surveying - general
Cartography - general
Data acquisition - platforms, instrumentation and equipment
- geodesy
- topographic and cadastral surveying
- hydrographic surveying
- photogrammetry and orthophotography
- digitizing and data structures
- statistical and cartographic techniques
- raster and vector techniques
image analysis
- geographic information systems
Interpretation and applications - general and perceptual
- geology and landforms
- soils
- snow and ice
- water
- atmosphere
- vegetation and ecology
- agricultural and land use
- thematic, planning
History of surveying and cartography
Atlases, sheet maps and map librarianship

The list above shows the contents page of *Geographical Abstracts G: Remote Sensing, Photogrammetry and Cartography*

The Vsesoyuzny Institut Nauchno-Tekhnicheskoi Informatsii (VINITI) [811] has published the monthly *Referativnyi Zhurnal Geodeziya i Aeros'emka* [294] since 1954. The journal can be consulted retrospectively as *Referativnyi Zhurnal Astronomiya i Geodeziya*. Several thousand informative abstracts covering the fields of photogrammetry and aerial surveying are published each year and are listed by author and subject in the annual index.

One of the most useful abstracting publications currently available is *Engineering Information. Technical Bulletin: Remote Sensing* [284]. The *Technical Bulletins* present informative abstracts drawn from over 4,500 articles and papers in journals, books, technical reports and conference proceedings from around the world on a comprehensive range of remote sensing topics. The subject areas covered range from civil and environmental engineering applications, through aerospace, electrical and electronic engineering, to instruments and measurements. There are 950 entries which are indexed by author, author affiliation and subject. Unfortunately this *Bulletin* is no longer published and will not be supplemented;

however, it may still be possible to obtain it via the inter-library loan system. Of further interest to the researcher concerned with fields related to remote sensing is *Engineering Information. Technical Bulletin: Satellites* [285]. The layout of this guide is similar to that for *Engineering Information. Technical Bulletin: Remote Sensing*, with broadly corresponding subject areas that are based on the classification used for other Engineering Index publications. The entries of interest are mainly concerned with the technological developments and plans of various international space research centres. Both of these publications are based on information filed in the COMPENDEX [310] database between January 1981 and February 1984. The COMPENDEX database contains over 1.5 million items and is a computer-readable version of the *Engineering Index Monthly* [282]. The *Engineering Index Monthly* covers the world's technological literature in all engineering disciplines, including remote sensing, and citations take the form of bibliographic records and abstracts drawn from journals, monographs, technical reports and conference proceedings, as well as some grey literature. The source publications that are scanned each year can be consulted in *PIE: Publications Indexed for Engineering*. Abstracts from the *Engineering Index Monthly*, an example of which is given below, are cumulated annually in *The Engineering Index Annual*, which features an alphabetic entry sequence by main subject headings and subheadings. *The Engineering Index Annual* also contains the *PIE: Publications Indexed for Engineering*, author and author affiliation indexes.

099416 SATELLITE REMOTE SENSING OF ATMOSPHERIC OPTICAL DEPTH SPECTRUM. An analysis of the atmospheric radiative transfer processes suggests that the diffuse radiance emerging at the top of the Earth's atmosphere in the position of a satellite can be approximated by a linear relationship with the optical depth. It is found that the variation of the optical depth is associated with the changing of aerosol size distribution. The optical depth increases with the presence of large aerosol particles. Thus, inhomogeneities in aerosol size distribution can affect the high-altitude atmospheric radiance in the visible wavelengths (400-800 nm). (Edited author abstract) 22 refs.

Aranuvachapun, Sasithorn (Univ of Exeter, Exeter, Engl). *Int J Remote Sens* v 7 n 4 Apr 1986 p 499-514.

NASA has published the *Abstract Newsletter: NASA Earth Resources Survey Programme* [275] since 1974 under several different titles. This weekly newsletter provides abstracts and full bibliographic citations of research and development reports from NASA agencies, contractors and grantees. Remote sensing and its environmental applications receive a comprehensive coverage. The information is also available online as part of the NTIS (National Technical Information Service Database) [318], which corresponds to several publications, including the journal *Government Report Announcements and Index* [235], which is fully described in the previous chapter. Copies of reports cited and abstracted in the *Abstract*

Newsletter: NASA Earth Resources Survey Programme can be obtained from the NTIS [799] in microfilm, microfiche, tape or hard-copy formats.

The American Institute of Aeronautics and Astronautics (AIAA) [747] has issued the bi-monthly *International Aerospace Abstracts* [289] with support from the Scientific and Technical Information Branch [771] at NASA since 1961. This journal contains over 40,000 abstracts each year that cover aeronautics, astronautics and the space sciences. The citations are drawn from journals, books, conference proceedings and monographs, although some of the articles are translations of foreign-language material. The machine-readable equivalent of this journal is operated by the AIAA as the AEROSPACE DATABASE [307] (called NASA in Europe). The database contains extracts from the abstracting periodicals *International Aerospace Abstracts* and *Scientific and Technical Aerospace Reports* [240]; the latter serial, which is a very comprehensive abstracting journal in its own right, covers the report literature on space and aeronautics that has been generated by NASA, NASA contractors, government agencies and research organizations. The AEROSPACE DATABASE is one of the key online systems for locating information on remote sensing and satellite technology, which reflects the bibliographic contribution made to the database by *International Aerospace Abstracts*. Original documents that are abstracted in the journal are available through the library service of the AIAA in photocopy or microfiche formats.

AESIS Quarterly [276], an indexing and abstracting periodical that has been available since 1976 from the Australian Mineral Foundation (AMF) [512], provides details of Australian literature, together with non-domestic publications that relate to Australia. Coverage is limited to the earth sciences, which are divided into seventy-eight subject headings, including remote sensing and cartography. The journal features author, locality, map, stratigraphic and well number indexes and an annual cumulation is available on microfiche. Of particular interest is *AESIS Special List No. 12: Remote Sensing and Photogeology, 1976 - June 1982* [277], which is updated periodically and produced from AESIS (Australian Earth Science Information System) [309], an online bibliographic database of over 30,000 records. The AMF has also produced the monthly current awareness bulletin, *Earth Science and Related Information: Selected Annotated Titles* [281] since 1983. The journal covers selected information from over 140 periodicals; each entry is usually briefly annotated and there are usually several items on remote sensing in each issue. *Earth Science and Related Information: Selected Annotated Titles* is computer-based and features automatically generated subject, locality, author, map sheet, mine/deposit/well name, stratigraphic and serial indexes. A bi-annual cumulation that collects titles under the broad subject headings is available on microfiche and several titles scanned for the database are listed in each January issue, although a list of these may also be obtained separately on request.

The Centre National de la Recherche Scientifique (CNRS) [602] publishes two abstracting periodicals that have a remote sensing interest: *Pascal Explore: Part 48. Environment Cosmique Terrestre, Astronomie et Geologie Extraterrestre* [291], and *Pascal Explore: Part 49. Météorologie* [292]. These periodicals have been published since 1972 under various names as different parts of *Bulletin*

Signalétique, which changed its name to the current title of *Pascal Explore* in 1986. The journal publications are based on entries in the PASCAL [319] database, a French- and English-language file with nearly 6 million records and an annual update of over 500,000 items. The file has been online since 1972 and may be searched via ESA-IRS and TELESYSTEMES-QUESTEL.

The major English abstracting service is provided by *Science Abstracts*, produced by the Institution of Electrical Engineers (IEE) [726]. The content of *Science Abstracts* is generated from the INSPEC [315] database and covers *Computer and Control Abstracts CCA* [279], *Electrical and Electronic Abstracts (EEA)* [282] and *Physics Abstracts (PA)* [293]. Each of these journals has a particular technological focus and all of them include abstracts from articles that represent the latest state-of-the-art developments in remote sensing. Each issue contains author, corporate author, conference proceedings, bibliographic and book indexes, together with a detailed subject guide. These are all cumulated in separate six-monthly index issues, and cumulative indexes to over 900,000 abstracts, covering the 1981-1984 period, are also available from the IEE. Inclusion of information in the respective abstract serial is dependent on the emphasis or subject of the paper under consideration. This is illustrated, together with a brief explanation as to why the articles are abstracted in a particular journal, in the following extracts:

21596 An approach to tree-classifier design based on hierarchical clustering. Wang Ru-Ye (Inst. of Remote Sensing Technol. & Application, Peking Univ., China).
Int. J. Remote Sens. (GB), vol.7, no.1, p.75-88 (Jan. 1986).
The tree classifier has proved to be an effective method of statistical pattern classification but the designing of the tree structure is not usually easy. In this paper, a new design approach based on hierarchical clustering is proposed. The design principle and the algorithm of the method, as well as the experimental results for both test data and remote-sensing data, are discussed. The design process of the method is simple and the performance of the classifier designed by the method has been proved to be satisfactory. (6 refs.)

CCA 86 21596 deals with the hierarchical clustering analysis of remote sensing data.

6146 Remote sensing. S.Murai (Inst. of Ind. Sci., Tokyo Univ., Japan).
J. Inst. Telev. Eng. Jpn. (Japan), vol.39, no.2, p.158-62 (Feb. 1985). In Japanese.
Reviews remote sensing activities and summarizes space station projects. The topical areas include solar observation, meteorological observation, oceanographic observation, Earth observation, geographic observation, and equipment necessary for space stations. (6 refs.) *K.B.*

EEA 86 6146 mentions equipment that is used in remote sensing.

57421 On the application of meteorological satellite imagery for monitoring the environment. G.E.Hunt (Centre for Remote Sensing & Atmos. Phys. Group, Imperial Coll. of Sci. & Technol., London, England).
Earth-Oriented Appl. Space Technol. (GB), vol.4, no.4, p.239-45 (1984).
Meteorological satellites now routinely monitor the Earth and provide measurements of the atmosphere, ocean and land surface which are used for studies of weather and climate. In this paper a review is given of current knowledge, with particular emphasis directed towards remote sensing of clouds, atmospheric energetics, mesoscale meteorology and the radiation budget. (32 refs.)

PA 86 57421 applies remote sensing techniques to atmospheric studies.

The *Applied Science and Technology Index* [278], which selects entries from 335 of the key English-language periodicals on science and technology, is a useful source of aerospace engineering information. The index is published monthly, with quarterly and annual cumulations; and although remote sensing is not indexed as a term, citations to the subject can be located under the headings of photogrammetry, mapping and aerial survey. A corresponding online version is available from 1983, and it currently contains over 150,000 records, with an average of 60,000 items being added per year.

Another useful publication is the weekly *Current Contents: Physical, Chemical and Earth Sciences* [280], a service (which reproduces and indexes the contents pages from journals and multiauthored books) that has been available from the ISI since 1961. The ISI also produces the *Science Citation Index* [295], a calendar-year compilation that is issued on a bi-monthly basis with an annual cumulation.

The *Science Citation Index* comes in three parts, the Source Index, Permuterm Subject Index and Citation Index, all of which may be used via extensive cross searching to identify pertinent publications from a very broad subject range. The *Science Citation Index* and *Current Contents* are available online as SCISEARCH, which indexes significant literature items from over 4,000 journals. The file currently contains over 7 million items and more than 600,000 are added each year. It is often advisable to use the online version for quick reference enquiries, as getting to grips with the main index is a fairly formidable task.

4.4 State-of-the-art Reviews

Remote sensing is not endowed with many review publications, which is a pity as reviews are probably the most time-saving and comprehensive introductions to a new subject that can be obtained. By definition, a review combines, but does not comment on, a variety of work on the same subject that has been published elsewhere. Most of the review series that deal with remote sensing and its related disciplines volunteer an evaluation or discussion of the topic under consideration, which lends a certain touch of originality to the work. Reviews that offer an interpretative commentary are very useful, although in analyses of highly technical work, critical evaluations or suggestions are of little use unless they are backed up by additional results from research.

It has been previously mentioned in this chapter that the references cited at the end of a review paper are useful bibliographic sources. The range of citations offered is obviously dependent on the scope and depth of a review, but in most cases this will be sufficient to offer an excellent retrospective overview of a particular topic.

The major review serial in this field of remote sensing is *Remote Sensing Reviews* [302], issued by Harwood Academic Publishers on an irregular basis since 1984. This serial provides a forum for international discussion on the acquisition, nature and processing of data for remote sensing, together with analysis of the research and operational practices of the application fields. The reviews focus on geology, oceanography, hydrology, cartography, resources survey, crops, soils, land use, limnology, climatology, forests, lands and planets. Researchers do not have to subscribe to the series and may purchase the individual issues that cover their field of interest. Each issue is multi-authored, has an overall editor, and includes extensive references and subject indexes.

The Technology Applications Centre (TAC) [743] has published *Remote Sensing of Natural Resources: A Quarterly Literature Review* [301] since 1973. This publication is perhaps more of a bibliographic literature update than a review, as it contains a collection of citations and abstracts from recent periodical and non-periodical publications. Remote sensing theory and applications are covered in the fields of geology, environmental quality, hydrology, vegetation, oceanography, regional planning and land use, data manipulation, instrumentation and geographic information systems. No cumulations are issued and each edition includes an author index, NASA technical briefs, news and a calendar of forthcoming events.

The Society for Photo-optical and Instrumentation Engineers (SPIE) [787] regularly holds conferences and workshops, the proceedings of which often contain reviews of specific themes in remote sensing. Of particular interest are *Remote sensing: a critical review of technology (SPIE Vol. 475)* [300] and *Digital image processing: a critical review of technology (SPIE Vol. 528)* [297]. The former title reviews the progress made in specific areas of remote sensing and discusses future prospects. *SPIE Vol. 528* is a collection of twenty-one papers that evaluate algorithms and concepts, implementation issues, application fields and miscellaneous applications. Also of interest is *Recent advances in civil space remote sensing (SPIE Vol. 481)* [299]. Very useful and extensive bibliographies are included in this proceedings volume, which reviews land observation remote sensing, meteorology, oceanography, agriculture and hydrology.

Advances in Space Research [27], the official journal of the Committee on Space Research (COSPAR) [604], regularly reviews the latest developments in space and remote sensing research. The collected papers that make up each issue include discussions and analysis of specific topics.

Nearly all of the regular serial publications on remote sensing include reviews. Books and conferences, together with organizations and current satellite technologies, are regularly reviewed in the excellent *EARSeL News* [349]. The journals *Mapping Sciences and Remote Sensing* [85], *International Journal of Remote Sensing* [68], *Remote Sensing of Environment* [104], *Rivista del Catasto dei Servizi Technici Erariari* [108], *Space* [114] and *Space Markets* [117] regularly publish review papers and articles which are all sources of useful literature citations.

The most useful general guide to reviews of a wide range of remote-sensing-related subjects is the *Index to Scientific Reviews* [298]. Published by the ISI [765], each semi-annual edition indexes over 25,000 review articles from 3,000 primary journals and periodic review series, in more than 100 subject disciplines. The layout is similar to all ISI publications, with research speciality front, source, permuterm subject and corporate indexes. Remote sensing, photogrammetry and related fields are extensively cited in the various indexes. The *Index to Scientific Reviews* is compiled from the SCISEARCH online database, which identifies key title words such as 'Advances', 'Review', etc., and selects articles with more than fifty references. An allied ISI publication, the *Science Citation Index* [295], identifies over 10,000 categorical, critical and narrative reviews with an 'R' (review) coding each year.

4.5 Online Databases

Over the last decade there has been a dramatic increase in the number and type of information sources that may be searched online. Machine-readable storage systems were originally conceived as an aid in the compilation of printed abstracting, bibliographic and indexing publications; however, advances in software, hardware and communications meant that the databases' information storage facility could be used to retrieve literature information from remote computer terminals in real time. It has become apparent that a reversal of the former situation has occurred and the majority of online databases now contain more current information than the abstracting and indexing publications that they generate.

The basic elements that make up an online information retrieval system are summarized in Figure 1. The database producer processes a variety of documents which are stored on a machine-readable database, from which abstracting and

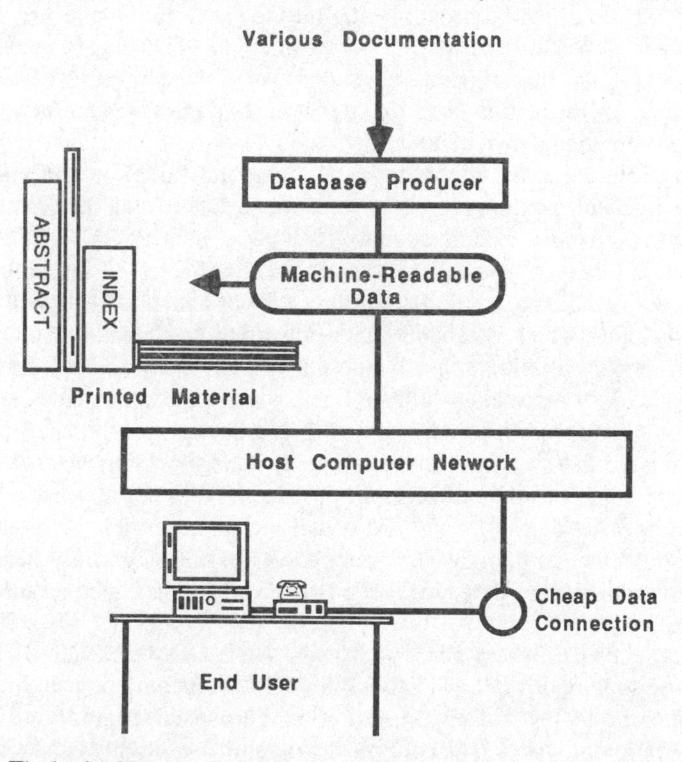

Figure 1 The basic elements of an online database network

indexing publications are generated. In some instances the database producer may also act as the host which is searched by the user; however, it is more often the case that the machine-readable data from the producer are entered onto a host-service operated by an organization with a computer network that offers several different database files. The ESA-IRS [498] is an excellent example of a large host, and offers over 35 million computer-readable bibliographic records in more than seventy separate databases. Other major hosts include BRS, DIALOG, DATASTAR and SDC/ORBIT in America and FIZ-TECHNIC, INKA and TELESYSTEMES-QUESTEL in Europe. Users contact the host via a tele-communication link and are then at liberty to interrogate the database of interest at their leisure.

There are many advantages in online access to large stores of bibliographic, full-text and factual data. A major benefit is rapid admission to a variety of files which are usually more current and complete than the published equivalents. In addition, the database is usually easier to search retrospectively than large library collections, which may not be available for browse purposes. There are also a lot of disadvantages, not least the expense of maintaining communications with a host and requiring skilled operators to access the databases. The various hosts use different software systems and the terminal operator must be conversant with the varying commands necessary for accessing a variety of host computers. There is also a need for the user to thoroughly define the subject area and use pertinent keyword combinations to retrieve relevant information. This may be regarded as a disadvantage by some investigators, but it does necessitate a degree of familiarity with the subject of interest, thus ensuring that users do not waste time and really know what information they require.

The database situation has recently become very favourable for the individual seeking literature information on remote sensing and its related disciplines. Prior to the advent of remote-sensing-specific databases such as RESORS (Remote Sensing Online Retrieval System) [320] and GEOBASE [311], researchers were wholly reliant on the larger multidisciplinary services such as the AEROSPACE DATABASE [307], which is called NASA in Europe, and the National Technical Information Service Bibliographic Database (NTIS) [318]. These services are excellent but do not represent wholly reliable methods of retrieving specialized remote sensing citations, which may be distributed through several such databases. It is to be stressed that the larger information services should always be consulted for a thorough search as they offer more documents on remote sensing than the more recent specific databases and are usually more accessible through remote terminals. However, as the new services evolve they will certainly become key sources, primarily because they have been conceived and designed around remote sensing.

The appearance of full-text files such as SPACE COMMERCE BULLETIN [325] and sections of SATELLITE NEWS [322] is encouraging and reflects a growing interest in the current situation in aerospace engineering and the commercialization of space. The continued availability of highly technical factual databanks like SATELDATA [321] and SPACECOMPS [326] suggests that online services have applications in many areas other than the retrieval of purely bibliographic information.

RESORS (Remote Sensing Online Retrieval System) [320] is a database operated by Gregory Geoscience Ltd., on behalf of the Canada Centre for Remote Sensing (CCRS) [548]. This information service provides rapid access to bibliographic references relating to the techniques, instrumentation and applications of remote sensing, photogrammetry and digital image processing. RESORS currently contains over 50,000 documents and is updated by 400 to 500 records each month. In 1986 nearly 220,000 literature citations, from over 9,000 information sources, were made available to remote sensing specialists. The document types include reports, journals, conference proceedings, theses, preprints, catalogues and patents. Nearly 3,000 of the reports are in French, and the database also stores documents in English, German, Spanish, Portuguese and Japanese.

In addition to bibliographic items, RESORS also offers SATLIS, a listing of past, present and future remote sensing satellites, and the ACRON file, a compilation of remote sensing acronyms. The RESORS Slide Collection, yet another facet of the database, consists of a set of over 5,000 slides with accompanying annotations, and is described in the section on audiovisual materials. Any literature that deals directly with earth observation from a distance, or the processing and analysis of such data, may be included on the database. The system is searched via keywords selected from a glossary of over 1,750 earth observation terms. Preferential weightings are applied to the keywords by the user, thus allowing a hierarchical search pattern. RESORS may be accessed through the DATAPAC system, which is available in Canada and through several other networks abroad. Enquiries may also be made via letter, telephone or telex and may be supported by a keyword list from the *RESORS Keyword Dictionary* [305], which is available free upon request. In order to make RESORS as comprehensive as possible, fully referenced copies of publications, that will be normally indexed from the full text using controlled subject keywords, may be sent for inclusion in the database. The entire RESORS file is also available on microfiche, although using this makes searching much more time-consuming. The most recent innovation with respect to this bibliography is its availability on a CD-ROM, which will be updated by the supplier twice each year. The CD-ROM will run on any IBM PC XT, AT, or compatible player.

The GEOBASE [311] database, which came online in 1987, is potentially of great interest to remote sensing specialists. Searchable through DIALOG, this system allows online retrieval of all Geo Abstracts Ltd. [724] publications, including the excellent abstracting journal, *Geographical Abstracts. Part G: Remote Sensing, Photogrammetry and Cartography* [287]. The GEOBASE service, together with RESORS, provide access to literature directly concerned with remote sensing, and these two services are representative of the dedicated and subject-specific databases that evolve around any technical subject.

The concept of specialized machine-readable stores of information has been realized by the ISPRS [501], which currently anticipates the establishment of an information retrieval system that will be accessed by postcard enquiry. According to Hothmer (1984), the database will offer access to bibliographic and factual information. It is envisaged that the subject coverage will be broad enough to include the interests of societies related to the ISPRS and estimates are that 5,000 documents drawn from conventional, factual and grey literature will be added to the database each year. It is further expected that compilations from other sources will be repackaged to fit into the database to give it a truly international coverage. A host has yet to be decided, but ESA-IRS and INKA are possibilities.

The Australian Mineral Foundation (AMF) [513] produces two databases, AESIS (Australian Earth Science Information System) [309] and ERISAT, a small computer-based literature index from which the monthly *Earth Science and Related Information: Selected Annotated Titles* [281] information bulletin is produced. The AESIS database contains over 30,000 geoscience documents relating to continental Australia. Remote sensing is amongst the wide range of disciplines covered and the database is particularly interesting as nearly half of the documents it contains are open-file company reports. AESIS is searchable online via AUSINET and a hard-copy equivalent is issued as *AESIS Quarterly* [276].

A series of databases are available for general information on space technology and science. Two of these, SATELLITE NEWS [322] and SPACE COMMERCE BULLETIN [325], are accessible through the host NewsNet. SATELLITE NEWS is a full-text file of peripheral interest to those researchers concerned purely with remote sensing; however, it does cover conferences, updates on NASA initiatives, programmes and policies, launches and satellite reports. A similar coverage is extended by SPACE COMMERCE BULLETIN, which has been available since 1984. Although this file concentrates on marketing, economics and policies, specialized topics such as the commercialization of remote sensing also receive detailed attention. Almost identical in name and subject coverage is a new database, SPACE COMMERCE ONLINE [324], which has only been available since 1985. Although full-text retrieval of documents relating to space law and policy is offered, this database differs in form from the previous systems, as it contains over 1 million bibliographic abstracts on the commercialization of space.

The ESA produces two useful databases that deal largely with the highly technical side of aerospace engineering and remote sensing. The SATELDATA (Satellite Databank) [321] file is produced by ESTEC [497] and contains over 1,000 records that give the precise parameters of various satellites and satellite-related equipment. Data are available on both current and future systems, including Meteosat and Landsat. SPACECOMPS [326] is a very similar database which has been available from ESRIN since 1970. Details of spacecraft components and their respective test reports, histories and uses may be retrieved online and the database contains over 11,000 reports and specifications which are updated as and when further information becomes available from the numerous space organizations that contribute to the databank. Similar in approach is the TECHNICAL REFERENCE FILE (TRF) [329], a subset of a database produced by the National Space Science Data Center (NSSDC) [775]. The TRF is a bibliographic database that contains published papers and reports on specific instruments that are present on a variety of spacecraft. The file is used to generate project bibliographies and to account for reports that may have been used to document spacecraft instrumentation. Over 36,000 documents are entered in the database and copies of some of the reports are held on microfiche at the NSSDC. A complete printout that runs to 476 pages and alphabetically lists all of the document titles available on the TRF may be obtained from the NSSDC if the user wishes to manually search the database.

A vast computer-retrievable store of information on cartographic-data products, aerial photographs, digital data and maps is available at the National Cartographic Information Centre (NCIC) [773], which is part of the USGS National Mapping Division. The NCIC database is a clearing-house for all cartographically related information in the U.S. and contains data that have been volunteered by Federal, State and local government agencies, as well as academic, commercial and industrial organizations and institutions. A similar initiative, although more global in outlook, is the establishment of the GRID (Global Resource Information Database) under the United Nations Environment Programme (UNEP). GRID is an extension of other UNEP databanks, and has the aim of disseminating environmental data to the international user community via a telecommunications network based on a complex computerized information service.

Apart from the databases that may be relied upon to return information of interest to remote sensing specialists, there are several general files that contain relevant information. One of the largest of these is the AEROSPACE DATABASE [307], which is produced by the AIAA [747] and NASA's Scientific and Technical Information Branch. The AEROSPACE DATABASE is available outside the United States as the NASA file, which is searchable through ESA-IRS. The database contains over 1,600,000 citations and nearly 6,000 items are entered into the file each month from its contributing serial publications, *International Aerospace Abstracts* [289] and *Scientific and Technical Aerospace Reports* [240]. All aspects of space science and technology are covered by the database, which is particularly useful for remote sensing literature retrieval. The AIAA library can provide most of the documents cited in the AEROSPACE DATABASE as photocopies or microfiche, and the library also offers a document delivery service from its comprehensive holdings of more than 30,000 books, 1,600 current journals, 750,000 NASA and AIAA microfiche and many thousands of conference proceedings and papers.

A very useful service is provided by the NTIS (National Technical Information Service Bibliographic Database) [318], whose coverage includes aerospace, atmospheric sciences, civil engineering, earth sciences and oceanography, electronics, electrical engineering and space technology. The NTIS database records research, development and engineering reports from U.S. Federal government agencies, their contractors and grantees. The file has over 1,200,000 entries and is the online version of the bi-weekly *Government Reports Announcements and Index* [235].

Searches from the NTIS database are available from the Published Search Service operated by Microinfo Ltd [728]. Several of these are on remote sensing and all contain full bibliographic citations with complete technical summaries. A

Published Search Catalog [304] on all subject areas is available from Microinfo, who also publish searches from COMPENDEX and several other files.

One of the largest database services is COMPENDEX [310], from Engineering Information Inc. Over 1,500,000 citations are entered on the database, which is updated by nearly 10,000 items each month. COMPENDEX is the machine-readable version of the *Engineering Index Monthly* [283], which covers the world's technological literature in all engineering disciplines. Documents on remote sensing, aerospace engineering and electrical and electronic engineering are all summarized and indexed on this database, which includes journals, reports, monographs and conference proceedings as its source publications. All of the documents that are retrieved online may be ordered via the terminal using the PrimorDial service.

The INSPEC [315] database contains over 2.3 million bibliographic records and is the most comprehensive English-language information service available for computing, electronic and electrical engineering, instrumentation systems, computer programming and computer applications. The Institution of Electrical Engineers (IEE) [726] generates *Science Abstracts* from INSPEC. Three files make up *Computer and Control Abstracts* [279], *Electrical and Electronic Abstracts* [282] and *Physics Abstracts* [293]. Remote sensing is included in the database and is referred to a specific file dependent on the technological slant of the individual paper. The database runs from 1969 and is available from a large number of hosts, including BRS, CAN/OLE, CEDOCAR, DATA-STAR, ESA-IRS, INKA and JICST.

The Institute of Electrical and Electronics Engineers has recently set up the online IEEE FINDING YOUR WAY [314] database. Although full details of the service are not yet available it appears to be very useful. The Tutorial Database File suggests monograph and review publications that should be read and courses and meetings that should be attended. A second file, the Catalog Database, is a comprehensive list of all IEEE publications that are available. Complete bibliographic, product and price information is displayed for any titles that are selected for viewing from either of these files. The database also offers a gateway into Videolog, an online news database and component selection guide. Over 650,000 electrical and electronic components from 700 manufacturers may be consulted via this non-IEEE service.

Several other online information services are of use for remote sensing literature retrieval, amongst these being PASCAL [319], the online equivalent of *Pascal Folio*. This database has over 6 million records and may be searched in the form of separate subject-specific files that correspond to the numerous printed equivalents. It is a particularly useful source of information on electrical engineering and computer science literature. Bibliographic information on surveying, mapping and remote sensing is found in the GEOLINE [312] database, which is produced by the Federal Institute for Geosciences and Natural Resources [612] and may be searched through INKA. The database covers geoscience literature from a wide range of publications in English, French, German, Russian and several other languages. The AGRIS [308] database is maintained by the FAO and may be searched through

DIMDI, ESA-IRS or FAO. The current file holds over 1.2 million bibliographic records that are mainly concerned with agriculture, but some remote sensing documents are also included.

Several other remote sensing organizations have small databases that are mainframe- or PC-based. Unfortunately external user access is not usually feasible on these stores, which are often difficult to identify and incompatible with other computer systems.

The number of online databases that are currently available is high, and others that cover different subject areas are appearing all the time. There are several online database directories that provide up-to-date information on the availability, size and facilities of these services. One of the most comprehensive is the *Directory of Online Databases* [303], published since 1979 on a continuing basis by Cuadra Associates. A similar directory, *Computer Readable Databases: A Directory and Data Sourcebook* [306], is available from Elsevier Science Publishers as a two-volume set, with the volume on Science, Technology and Medicine being most relevant to remote sensing. One problem with these guides is that they are generally out of date by the time of publication, although one way of keeping up to date with them is to consult the online versions. The Cuadra publication is available as the DIRECTORY OF ONLINE DATABASES, which can be searched on Data Archiv AB, DATA-STAR, TELESYSTEMES-QUESTEL and WESTLAW, and gives the same sort of information (names, number of records, update frequency, subject coverage, details of online processors etc.) as the printed equivalents. Over 3,000 publicly available databases worldwide are recorded in this file, which is updated twice a year. The *Computer Readable Databases: A Directory and Data Sourcebook* directories are also available online as the DATABASE OF DATABASES, which may be searched on DIALOG. Coverage extends to over 2,500 databases from all subject disciplines worldwide and most of the information on this file is from database producers, although additional data are also recorded.

4.7 Newsletters and Bulletins

Newsletters are some of the most important publications and means of keeping up to date in remote sensing. Although usually brief in content and often irregular in issue, their topicality and relevance cannot be denied and a list of over seventy newsletters, which are judged to be the most important in remote sensing, is given in Part II [330-401]. Far more than this exist, but it has been the author's general policy to exclude those of irregular issue, such as the somewhat irregular bulletins emanating from universities and some other organizations.

In remote sensing there appear to be three kinds of newsletter: newsbrief types of publication that provide general overviews of remote sensing activity; more specific bulletins that contain reports on developments within the issuing organization as well as news; and the mini-journal, offering technical papers and bibliographic, abstracting and current awareness services. The news coverage of most bulletins is wide, embracing the international remote sensing community and reporting on policies, politics, legislation, funding, technical and research developments, new organizations, project progress and services. Most newsletters also carry advance conference notices in the form of a diary or calendar, together with reports on recent meetings, workshops and symposia, product or service advertisments and book and publication reviews. Only a few of the more select publications feature the technical papers that have been mentioned; when they do, it can cause problems as newsletters are infrequently subject to cataloguing procedures by libraries and therefore suffer from poor bibliographic control. This lack of control usually means that a technical paper appearing in one of these issues may be more or less lost as it will not find its way into the available abstracting, indexing and online services.

Of the North American newsletters, the *Washington Remote Sensing Letter* [400] provides the best general news coverage. Published twenty-two times per year from the National Press Building in Washington [802], this excellent, closely typed newssheet acts as forum for the dissemination of news on many topics. More normal inclusions are book reviews and conference notices, but there is also detailed information on policies, politics, legislation and government, federal agencies and commercial companies' research activities in remote sensing and space technology. Three new notes issues come from EOSAT [741]; one of these, *EOSAT Landsat Data Users Notes* [351], is a continuation of *Landsat Data Users Notes*, previously issued by the EROS Data Center [742]. The National Oceanic and Atmospheric Administration (NOAA) [774] has transferred the entire *Landsat Data Users Notes* collection to microfiche. Five fiche make up the set, all of the original colour reproductions are retained, and a useful index has been appended to facilitate the location of information. The glossy A4 size sheets included in the new series are designed to be ring-bound and contain news on EOSAT purchasing policy, receiving station status, government views and data applications. The second EOSAT publication is a wholly application-oriented newsletter, *EOSAT Landsat Application Notes* [350]. Each issue concentrates on a particular application and often features well reproduced colour images illustrating that application, together with a selected reading list. *EOSAT Landsat Technical Notes* [352] are published on an irregular basis as an attempt to maintain user awareness in the developments in Landsat data processing and technology. The *Notes* series are useful publications which complement each other, while at the same time acquainting customers with the potential commercial applications of remote sensing. The NASA/Goddard Space Flight Center [762] publishes the *National Space Science Data Center Newsletter* [373] on a bi-monthly basis. The fact that this publication originates from within a government agency is reflected in the topical news features on space policy and details of advances in official research projects. The *National Space Science Data Center Newsletter* also contains useful technical data on both online and offline computer retrieval systems being developed at the Center.

The Geosat Committee Inc. [761] publishes *Geosat News* [359] on an irregular basis. Apart from news on topics including political developments in remote sensing, this bulletin considers committee news and projects, includes occasional technical articles and provides extensive advance notices of forthcoming events and publications. All of the previous newsletters contrast with the lighthearted and entertaining, but nonetheless informative, *Photogrammetric Coyote* [379]. Published as a quarterly by E. Coyote Enterprises Inc. [758], this large newspaper presents news, advertisements, equipment lists and tests in a very readable format.

Another useful newsletter publication is *Remote Sensing in Canada* [385], published on a quarterly basis by the Canada Centre for Remote Sensing (CCRS) [539]. Printed back-to-back in English and French, each issue features news items on Canadian remote sensing projects and comprehensive reviews and advance notices of forthcoming and past meetings.

In the U.K., the Remote Sensing Society's [753] members receive the free *Remote Sensing Society's News and Letters* [387]. This mini-journal is reprinted as a bi-monthly publication of Taylor & Francis Ltd., from the *International Journal of Remote Sensing* [68]. In addition to the comprehensive international news section of the parent journal, it contains Remote Sensing Society news. The three to four short technical letters included in each issue are intended to be forerunners and advance notices of research in progress; however, they are often papers in their own right. A more conventional publication, the *National Remote Sensing Centre Newsletter* [371], is produced on a quarterly, but often more irregular, basis by the National Remote Sensing Centre (NRSC) at Farnborough [703]. This news bulletin is very well put together and carries detailed information on NRSC, U.K. and foreign developments and activities in all aspects of remote sensing. The NRSC also produces a comprehensive series of remote sensing fact sheets that are available upon request. These double-sided A4 handouts feature excellently reproduced colour photographs accompanied by a brief descriptive narrative, and in appropriate cases, technical specifications. The sheets' subject coverage is very wide, embracing a great many applications of remote sensing, as well as describing past and present spaceborne imaging systems.

Apart from the other benefits of being a member of EARSeL [494], it would be worth becoming affiliated to the organization for its quarterly bulletin *EARSeL News* [349]. The regular features in this mini-journal include Association news and reports of Bureau and Council meetings, Members' news and discussion topics, recently published information on European and U.S. receiving station status and progress with new satellites, a current awareness service that extracts items from international publications, industrial news and more general information such as book, publication and conference reviews, together with detailed analysis of subjects due to be covered in forthcoming meetings. *EARSeL News* has a readership of over 1,000 scientists involved in remote sensing and its scope and content make it one of the most informative publications produced in Europe.

SPOT Image [598] publish the *SPOT Newsletter* [394] in English and French twice a year. Each issue is largely concerned with marketing and the commercial applications of SPOT data. A very useful mini-directory that reports on the current state of SPOT distributors and receiving stations is sometimes included, together with well printed examples of SPOT imagery. This newsletter has a sister publication with much the same content, namely *SPOTLIGHT* [393], issued on a quarterly basis by the SPOT Image Corporation [789] in the United States. The French organization Prospace [599] publishes its informative journal-style bulletin *News from Prospace* [375] about three times a year. Prospace represents French remote sensing interests and its involvement in this field is reflected in the journal's content, which includes the latest information on launchers, facilities and space programmes in France, together with the products and services offered by members of the Prospace organizations. The various divisions of the ESA [495, 496, 497, 498] produce several newsletters, including *Earth Observation Quarterly* [347], *ESA Bulletin* [353], *ESA Newsletter* [354], *Columbus Logbook* [342] and *News & Views* [374]. The most notable and widely circulated of these is *Earth Observation Quarterly*, which is published on a quarterly basis by the ESA Publications Division [497]. The content of this glossy newsletter is dominantly concerned with activities and developments within the ESA Earth Observation Preparatory Programme (EOPP).

Several very good newsletter publications are available outside Europe and the United States. Two of these are issued from the same place, the Interdisciplinary Center for Technology Analysis and Forecasting (ICTAF) [635], in Israel. The *Israel Remote Sensing Information Bulletin* [363] provides information on courses, workshops and meetings held by the ICTAF, as well as international news and several Israeli-authored technical papers in each issue. Every edition also features the 'ICTAF Bibliography', which is essentially a current awareness section that consists of photocopies from the contents pages of major international remote sensing journals. The 'ICTAF Bibliography' is also featured in the Center's bi-monthly *Space Research and Technical Information Bulletin* [391]; however, here the information is more complete and publications which have been abstracted and indexed are complemented by citations that have been recorded from online databases. The news section of this mini-journal is also very good and there are detailed reports on projects and activities that have been initiated by ESA, NASA and other aerospace and remote sensing organizations.

Short technical papers on remote sensing, as well as a calendar and map of Landsat flights across Australia, are features of the bi-monthly *Australian Landsat Station Newsletter* [334]. News reports outline the status of Australian and foreign remote sensing stations and in addition to the normal notices and reviews, a complete list of Australian Landsat Station (ALS) [515] reference centres is provided.

The Asian Association on Remote Sensing (AARS) [504] has published the quarterly newsletter *Asian Association on Remote Sensing* [333] since 1983. News and notices are included and the newsletter is distributed free to members of AARS and to individuals from interested non-member countries. A similar publication is the *ARRSTC Newsletter* [332], which is issued on a quarterly basis by the Asian Institute of Technology (AIT) [701]. In contrast to the previous newsletter, this bulletin reports on student affairs, progress and projects, as well as more general news. Another newsbrief, which originates within Thailand, is the *TRSC Newsletter* [399], a quarterly publication of the Thailand Remote Sensing Centre (TRSC) [700], available in English or Thai.

Newsletters provide very topical information that can be of great interest to one's research area, in a form that is cheap, or often free. Very few other publications have their immediacy and only occasionally report on other organizations' activities so completely. Many individuals and groups will receive newsletters from organizations they have dealt with as a matter of course; however, a request to be included on a newsletter mailing list will seldom be denied.

4.7 Regular Conferences on Remote Sensing

Conferences are probably the most useful means of keeping up to date with developments, techniques, applications and theories in remote sensing. Attendance at a conference may have several aims: to assess current state-of-the-art thinking in individual specialist fields; to present a paper or poster session on one's own research; or increasingly, to discuss business and either buy or sell expertise and products. Most of the major remote sensing conferences have exhibitors who have paid a fee to the organizing body for being allowed to promote their product or advertise their expertise. Regular attendees are the large commercial vendors such as SPOT Image [598] and EOSAT [741], together with firms that produce digital image processing and geographic information system hardware and software.

Two of the major considerations with regard to meetings are how to obtain the printed conference proceedings and identify papers of potential interest. This involves the use of indexing directories, and those concerned with conference literature have been detailed in the preceding chapter. The researcher or potential attendee will also need information on exactly when and where a conference, meeting or workshop is to take place; however, for an individual from an academic, industrial or commercial organization who is a recognized authority on remote sensing there should be no problem. Most groups and individuals that regularly publish the results of their research in remote sensing will be forewarned of conferences and asked to submit a paper several months in advance of the date upon which it is to be presented. For those who are not subject to such notice, there are a number of announcement sources of interest. Two very useful publications produced by the World Meetings Information Center are *World Meetings: United States and Canada* [249] and *World Meetings: outside the United States and Canada* [248]. Published on a quarterly basis by Macmillan Co. Inc., each of these journals is indexed by location, organization, subject and title. The journals and newsletters that are issued by various aerospace and remote sensing organizations invariably have diaries, calendars, or some other advance notice of conferences, symposia, meetings and workshops. A more recent announcement service is the online FAIRBASE-FAIR directory, which may be searched on DATA-STAR. This database came online in 1986 and has over 5,000 records of worldwide multi-disciplinary meetings and events that will take place up until the year 2000. The file is current and the updates, which occur about eight times per year, feature information on meeting organizers, location, dates and subject. A similar service is offered by the file INKA-CONF, which is available from the German INKA host; this database specializes in science, technology, aeronautics, space research and physics. The current file contains over 25,000 records and is updated on a weekly basis with information from journals, programmes, circulars and any other media that advertise conferences.

Conferences that are of interest to specialists in remote sensing and related fields are listed in Part II [415-442], and although announcements of theme, location and date are often given a year in advance of the meeting, it would be unrealistic to list these here. The reader is advised to contact the address of the organizing or sponsoring body for current information. The rest of this section provides a brief description of the sponsors and themes of major conferences in remote sensing and allied fields.

It is not easy to identify regular conferences and compare their relative merits, as many meetings are either specific to one aspect of remote sensing and related fields, or are more broadly based, encompassing a wide spectrum of interest. The former meetings generally have a small but highly specialized attendance, while the latter are often more of a social occasion than a scientific meeting, although the presented work is still of the highest technical merit. The preparation of meetings requires considerable effort and it is common for commercial and industrial organizations, societies, universities and government agencies to co-sponsor and organize conferences.

The ISPRS [501] holds a major Quadrennial Congress [433] that consists of papers selected by its seven Technical Commissions during their four-year period of operation. Proceedings are published as the *International Archives of Photogrammetry and Remote Sensing*, which replaced the *International Archives of Photogrammetry*, a proceedings series that dealt more exclusively with photogrammetry and aerial surveying, in 1980. Each Technical Commission is sponsored by a member organization, which organizes several Working Groups that are responsible for areas of research within the broadly defined field of a particular Technical Commission. Each Commission also arranges a Symposium [434], that may occur at any time within the four-year gap between Congresses. The proceedings of the individual Symposia are published by the sponsoring member as volumes of the *International Archives of Photogrammetry and Remote Sensing*, which reflect the current international status of research in remote sensing and are useful contemporary and retrospective records of scientific developments in these fields.

The main organization responsible for holding conferences, meetings and workshops on remote sensing and photogrammetry in North America is the American Society of Photogrammetry and Remote Sensing (ASPRS) [740]. The Society holds two week-long meetings each year: the Annual Convention [415], which is co-sponsored by the American Congress on Surveying and Mapping (ACSM) [745], and the Fall Convention [416]. Over 600 papers on all aspects of remote sensing are generally presented in concurrent sessions at these meetings. The proceedings have been published for the last twenty years as *Technical Papers from the Annual Meeting* and *Technical Papers from the Fall Convention*, and back issues are available from the ASPRS, which publishes an *Index to Technical Papers from Annual and Fall Meetings of ASPRS & ACSM 1975-1982 & 1983*

[246] that serves as a guide to thirty-three printed proceedings volumes. The ASPRS also holds specialized conferences on a number of themes, the most regularly occurring of which is the Biennial Workshop on Color Aerial Photography in the Plant Sciences and Related Fields [417], which addresses the methodologies and applications of aerial photographic interpretation in agriculture and plant science.

The Laboratory for Applications of Remote Sensing (LARS) [768] holds the annual International Symposium on Machine Processing of Remotely Sensed Data [436] at Purdue University. Both theoretical and application-oriented papers that are specific to data processing, digital image analysis and geographic information systems are presented, with most of the contributions coming from researchers in the United States. This conference series has been held since 1974 and the printed proceedings, *Machine Processing of Remotely Sensed Data*, are very informative and should appeal to anyone actively engaged in remote sensing research.

Established as a tribute to Willian T. Pecora, a 'founding father' of the LANDSAT system, the Pecora Symposium [442] has been held annually since 1975 in various cities of the United States. The Symposium, which is sponsored by the USGS and NASA, considers recent developments in space remote sensing and related disciplines. Presentations are from leading individuals in commercial organizations, industry, government and Federal agencies.

The International Symposium on Remote Sensing of the Environment [424] is held annually by the Environmental Research Institute of Michigan (ERIM) [759]. A very broad range of topics in remote sensing are considered and when it is not actually occurring at ERIM, the conference is co-sponsored by an organization from the country in which it takes place. The Symposium proceedings form an excellent retrospective archive that reflects developments in remote sensing, and they may be searched via a useful two-volume *Index and Abstracts* [243] publication that covers twenty-eight proceedings volumes for the period 1962-1980. The size of the symposium, allied with a requirement for state-of-the-art reviews and research on specialized applications, has led to a number of smaller thematic conferences being run in tandem with the central event. The most notable concurrent session that has emerged is the annual Thematic Conference on Remote Sensing for Environmental Geology [425]. Another conference at which research and development papers on the geological applications of remote sensing are presented is the Geosat Committee Inc.'s Geosat Workshop: Frontiers of Geological Remote Sensing from Space [426]. The papers featured in the proceedings of this meeting cover a wide spectrum of geologic interest and there are contributions from each of the six Geosat working groups.

The IEEE Inc. is a major sponsor and organizer of a variety of international meetings [428-432] that feature papers contributed by its wide membership, as well as by external research bodies. Probably the most interesting meeting to those in remote sensing is the annual IEEE Geoscience and Remote Sensing Society Symposium [432]. The Symposium is organized by the IEEE Geoscience and Remote Sensing Society [763] and features papers on a range of mainly technical themes. The proceedings are published in the *IGARSS Digest*, and more selectively in the official journal of the IEEE Geoscience and Remote Sensing Society, *IEEE Transactions on Geoscience and Remote Sensing* [61].

Other North American organizations that hold regular conferences are the Society for Photo-optical and Instrumentation Engineers (SPIE) [787], the Society of Photographic Scientists and Engineers (SPSE) [788] and the American Society of Civil Engineers (ASCE) [748]. In addition to the Annual International Symposium [440], the SPIE holds nearly fifty speciality meetings a year on various topics in electro-optical, optical and opto-electronic instrumentation and systems. Reviews of conferences on remote sensing and digital image processing techniques are occasionally published, together with the results of other meetings, as *Proceedings of the SPIE*. The Annual SPSE Conference [441] deals with engineering and technology advancements in imaging technology, and five to six other meetings are held each year on areas of specific interest to the SPSE's sixteen working divisions. The Engineering Surveying Division of the ASCE occasionally presents technical papers on photogrammetry, remote sensing, or aerial surveying related to engineering problems, at its Spring [419] and Annual Conventions [418].

The only major North American meeting held outside the United States is the Canadian Symposium on Remote Sensing [420]. The Symposium, which has been held every eighteen months for the last fourteen years, attracts an international audience, although most of the presentations detail the environmental applications of remote sensing within Canada. Each provincial centre and co-ordinating body for Canadian remote sensing activities usually makes a contribution to the Symposium and the results are published as the *Proceedings of the Canadian Symposium on Remote Sensing*. Another important Canadian event is the annual Canadian Advisory Committee on Remote Sensing (CACRS) meeting. Attendees usually include chairpersons of the CACRS working groups, provincial and national remote sensing society members, provincial representatives, and invited speakers from commercial and industrial organizations that specialize in marketing remote sensing products or expertise. The results of the various discussions are published as the *Canadian Advisory Committee on Remote Sensing. Annual Minutes* [338], which are an interesting guide to the policies and thinking of a highly organized committee and membership.

In the U.K. the Remote Sensing Society (RSS) [733] co-sponsors an Annual Conference [437] that deals with a specific theme in remote sensing. Proceedings volumes are usually available at registration, and back issues, several of which are available only on microfiche, may be obtained from the Society. The papers presented at these meetings are selected from an international pool of contributors, and are often of a broader nature than the conference title would suggest in itself. The Institution of Electrical Engineers (IEE) [726] holds numerous meetings on topics such as electronic and electrical engineering, computing, systems and instrumentation each year. For state-of-the-art information on technical developments in these and other fields, the International Conference on Antennas and Propagation [427] is very useful. As in many of the IEEE meetings, remote sensing is not always dealt with directly, but many of the contributions are relevant to image processing and aspects of microwave theory.

Elsewhere in Europe, EARSeL [494] organizes a series of workshops, meetings and symposia with financial aid from the ESA, the Council of Europe and the Commission of the European Communities. The most regular of these is the annual EARSeL/ESA Symposium [423], which is usually co-sponsored by the ESA. These meetings are well organized, bringing together representatives from most of the member laboratories for discussions that include EARSeL policies and recommendations, as well as contributions on a range of practical themes covered by the working groups of EARSeL. The International Cartographic Association [507] regularly runs a technical conference, Auto-Carto [435], that deals with the latest developments in spatial information collection, extraction, analysis and processing methodologies and applications. Each meeting is held in a different location and is usually co-sponsored by one or more organizations. A number of fields of activity, including topographical and cadastral surveying, satellite remote sensing, hydrology and digital cartographic information systems, are covered. Several other meetings are held or sponsored by European organizations, but many of these are very irregular, and some, such as the International Symposium on Acoustic Remote Sensing of the Atmosphere and Oceans [422], sponsored by the Consiglio Nazionale delle Richerche [636], are relatively new but gradually becoming more established.

It is more difficult to identify meetings that take place outside Europe and North America, but they are numerous, and several of the workshops and symposia, including the Congreso Nacional de Fotogrametría, Fotointerpretación y Geodesia [421], the Australasian Remote Sensing Conference [438], the Annual Information Seminar [439] and the Asian Conference on Remote Sensing, are well organized and useful sources of information. The latter event is held by the Asian Association on Remote Sensing (AARS) [504] in a different member country each year. It is particularly useful as it presents a forum for the presentation of material from these countries that may not be available elsewhere.

Reference

Hothmer, J. (1984). The information retrieval system ISPRS-IRS for literature and factual data. *International Archives of Photogrammetry and Remote Sensing* XXV (A6): 155-169

5 Language Problems in Remote Sensing

If the literature of remote sensing is consulted, it becomes obvious that papers and articles originally written in languages other than English receive little attention. A widely held belief is that foreign-language material is insignificant and that there are sufficient publications on the same topics in English to warrant its exclusion. This presumption is rather simplistic, especially in remote sensing, where the publications of developing and other countries are very important and can reflect the current state-of-the-art, or progress within that country, in a certain area.

5.1 Solving the Problems

The problems of obtaining translations of engineering literature are very well outlined by Glover (1985) and Wood (1976). The amount of translation activity for remote sensing is a great problem and there are few regularly published translation series. The only remote sensing periodical with a cover-to-cover translation into English is the Russian journal *Issledovanie Zemli iz Kosmosa* [70], a bi-monthly publication from the Akademiia Nauk S.S.S.R. [805]. The translation, which is issued by Harwood Academic Publishers as the *Soviet Journal of Remote Sensing* [113], features technical papers and abstracts of forthcoming articles. The only other serial publication that regularly features translations of Russian research papers is the quarterly journal *Mapping Sciences and Remote Sensing* [85]. Published by V. H. Winston and Sons Inc., with co-operation from the ASPRS [740] and ACSM [745], this journal has been issued since 1962 under the former titles of *Geodesy, Mapping and Photogrammetry* and *Geodesy and Aerophotography*. Each issue features direct translations of articles appearing in over ten well-established Soviet and Eastern European periodicals on geography, aerial photography, photogrammetry and remote sensing. State-of-the-art reviews, polemics, special features and translations of manuscripts and papers that are forthcoming from other sources are included. It is common for periodical publications such as *Vermessungstechnik* [124] to have either abstracts, or in the case of the French journal *Photointerprétation* [102], descriptive narratives, in several languages. Similarly, many of the journals issued by non-English-speaking countries may contain contributions in several languages; for example *Revista Cartográfica* [106] features papers in English, French, Spanish and Portuguese.

Russian-language material appears to be added as a priority to certain online bibliographic files, and in a zoom and frequency analysis of the INSPEC [315] and COMPENDEX [310] databases, the *Soviet Journal of Remote Sensing* and its parent publication *Issledovanie Zemli iz Kosmosa* were joint first in INSPEC and sixth in COMPENDEX, which also features a Soviet reference journal as its eighth entry. An edited version of the ratings, frequencies, and the time periods over which they were added to the database is listed in Table 1, and reveals an apparent interest in Russian remote sensing literature. Most online databases have quite significant amounts of foreign-language material in them and according to Large (1983), records from papers published in 1980 and added to the INSPEC database were eighteen percent foreign-language. Nedic (1982) found similar results when investigating citations in *Engineering Index Monthly* [283]: approximately 32,000 records, or twenty-eight percent of the papers recorded annually, were in a foreign language.

Table 1 Journal frequency analysis

File: INSPEC (1971-86)

1 91 ISSLED ZEMLI IZ KOSMOSA USSR
2. 91 SOV J REMOTE SENSING SWITZERLAND
3. 73 IGARSS 84 ESA SP 215
4. 52 ELEVENTH INT SYMP MACH PROC
5. 51 INT J REMOTE SENS GB
6. 49 ADV SPACE RES GB
7. 48 APPL OPT USA
8. 33 PROC SPIE INT SOC OPT ENG USA

COMPENDEX (1973-86)

1. 69 INT J RS
2. 63 RS ENVIRONMENT
3. 42 PHOTOGRAMM RS
4. 26 RADIO SCI
5. 14 J GEOPHYS RES
6. 12 SOV J RS
7. 9 IEEE TRANS GEOSCI RS
8. 8 IZV AKAD NAUK SSSR FIZ ATMOS OKEANA

The monthly *World Transindex* [414] lists translations that are deposited with the International Translations Centre (ITC) [508] in Delft. The ITC has been active since 1961 and receives contributions from an international network of centres, including the British Library Document Supply Centre (BLDSC) [713]. As well as publication of the *World Transindex*, the ITC's duties include the dissemination of scientific and technical translations to the user community, together with maintenance of the WTI (WORLD TRANSINDEX) database in co-operation with the Centre National de la Recherche Scientifique (CNRS) [602]. This file, the machine-readable version of the *World Transindex*, contains over 200,000 records which date back to 1978. Over 25,000 items are added each year from serial and non-serial publications and these can be ordered at the terminal via the PrimorDial Service of the host, ESA-IRS. The database announces translations of literature from all scientific and technical disciplines into Western languages from Adriatic and Eastern European languages, together with translations of Western languages into French, Portuguese and Spanish. The ITC runs an information service for all translations that have not been added to these sources and also publishes several retrospective indexes of translations that are available from the centre.

The BLDSC issues the indexing journal *British Reports, Translations and Theses* [3] every month. This publication concentrates on 'grey' or 'semi-published' literature that is not normally catalogued. Translations from British government organizations, learned institutes and societies are listed and there is also coverage of unpublished translations from the Republic of Ireland. Each edition features a keyterm index, and the quarterly microfiche cumulation is indexed by author, report number and keyterm. Remote sensing occurs only occasionally as a term, but translations on this and related disciplines may be located by browsing through the subject headings of Aeronautics, Earth Sciences, Electronic and Electrical Engineering, Computer Science and Space Technology. Records from this publication, together with literature contributions from a consortium of European data centres, are held in the SIGLE (System for Information on Grey Literature in Europe) [323] database. This file contains over 60,000 items, but records of translations only appear in SIGLE if they are translations of reports. *Current Serials Received* [263] is published by the BLDSC on an annual basis and in addition to cataloguing the 70,000 serial titles currently taken by the BLDSC, this document lists all of the cover-to-cover journal translations that are received.

Some large aerospace and remote sensing organizations, such as the Institut für Angewandte Geodäsie [610], may have their own 'in-house' translation services. The material recorded by these will usually be project-oriented and may be published as an occasional series, such as *NASA Technical Translations* [409], *U.S.S.R. Report: EARTH SCIENCES. Abstracts only* [413] and *Catalogue of NAL Technical Translations* [403]. The central translation indexing services of the ITC have already been described and they mainly cover European literature, but organizations such as the BLDSC, Aslib [709] and the Translations Research Institute [790] hold current translation indexes for the U.K. and United States.

If no translation of a work is available then the individual or group may consider some of the specialist translation services that are available. Lists of these can usually be obtained from the centres mentioned above; however, it should be noted that translation rates are high and can range from $40 per 1,000 words for European languages up to $100 per 1,000 words for less common languages. For this reason it is unlikely that most researchers will use translation services, unless they are lucky enough to be backed by substantial funding. Some idea of an article's content may be gleaned from using the dictionaries listed below as translation aids, but however useful these are, they are only suitable for occasional words and as such will not present information in context.

5.2 Aids to Translation: Interlingual Dictionaries in Remote Sensing

The most recently published specialist dictionary for remote sensing is the *Multilingual dictionary of remote sensing and photogrammetry* [411]. This authoritative work, which was issued by the ASPRS [740] in 1984, contains 1,716 sequentially numbered terms that are defined in English and translated into French, German, Italian, Portuguese, Russian and Spanish. The terms in each language index are alphabetically listed and cross-referenced to the English definitions. Useful sections in the dictionary list abbreviations and acronyms, and a selected bibliography is also included.

A 300 page working document, the *French-English glossary on SPOT, remote sensing and their applications* [405], was published by its author, S. Dyson, in late 1986. The glossary features over 3,500 entries, ten mini-glossaries concerned with applications, source document referencing, acronyms and diagrams. This large-format book has room for annotations and updates, and is also available on 3.5 inch 400 K or 800 K Macintosh/Word discs. An older glossary of space terms, *Recueil de terminologie spatiale* [412], was published by the European Space Agency in 1982. Another French publication, *Dictionnaire de télédétection aérospatiale* [410], was issued by Masson in 1982. Terms and their definitions are defined in French, with introductory material and indexes in both English and French. A *Dictionary of remote sensing terms, English-Spanish, Spanish-English* [404], which was published in Lima in 1980, is the only dictionary dedicated solely to English-Spanish translation of which the author is aware.

The *English-Russian dictionary on remote sensing of the natural earth resources* [407] was published in 1977 by the Caspij Publishing House and contains a straight list of 4,000 terms in English and Russian. The Editura Technia (Technical Publishing House) in Budapest published the *Dictionar poliglot di geodezie fotogrammetrie di cartografie* [402] in 1976. This is a very comprehensive work, with over 4,700 entries defined in English and included in separate German, French, Romanian and Russian sections. With over 10,600 selected terms used in space technology and an index of English and referred definitions, the *Russian-English space technology dictionary* [408], published in 1980 by Pergamon Press, is a dated, but useful reference aid for aerospace terminology. A similarly dated, but very valuable, translation aid is the ISPRS [501] 1969 *Multilingual dictionary for photogrammetry* [406]. This dictionary is divided into seven volumes, one each for English, French, German, Italian, Polish, Spanish and Swedish. Many of its 4,259 terms are concerned with photogrammetry, but it ranges across the whole field of remote sensing and each term may be cross-referenced to a particular language in any of the companion volumes.

According to Linding (1984) a dictionary, which should be completed by 1988, is currently under preparation by the ISPRS. This multilingual publication is expected to have around 3,000 entries in Arabic, Chinese, German, English, French, Greek, Hindi, Japanese, Portuguese, Russian, Spanish, Thai, Bengali and Turkish. A preliminary English glossary of entries and definitions of 3,000 items in photogrammetry, that provides an indication of the expected coverage and may be of use to some researchers, has already been published by Wolf (1980).

References

Glover, W. (1985). 'Translations.' In L. J. Anthony. *Information sources in engineering*. London: Butterworths.

Large, J. A. (1983). *The foreign language barrier*. London: André Deutsch.

Linding, G. (1984). Status of multilingual dictionary (ISPRS). *International Archives of Photogrammetry and Remote Sensing* XXV (A6): 199-208.

Nedic, Z. and McCoy, B. S. (1983). EI's inside look at technical translation. *Science and Technology Libraries* **3** (2)

Wolf, P. R. (1980). Tri-lingual glossary of photogrammetry terms. *International Archives of Photogrammetry* **23**-1310: 199-311

Wood, D. N. (1976). 'Translations.' In Mildren, K. W., ed. *Use of engineering literature*. London: Butterworths.

6 Sources of Remotely Sensed Imagery

6.1 Land Remote Sensing Satellite Data and Aerial Photography

Despite the widespread use of remotely sensed data, a comprehensive listing of the types, coverage, availability and quantities of those data has yet to be compiled. The archival control of the more popular forms of satellite imagery is very good, with prospective users usually being able to obtain current or retrospective material for their area of interest from one of the large commercial or governmental organizations. A considerable amount of potentially valuable imagery is, however, lost to some extent, because of both poor record keeping and poor archival practices. Although this affects satellite imagery, it is much more of a problem with aerial photographs, and tens of thousands of these are inaccessible to the potential user because the various government departments, universities and other organizations that have them are unaware of their value and do not even have up-to-date records of what material they may have. Nevertheless, it can be worth approaching the local councils, embassies, libraries and land survey organizations of a country for which imagery is required. Many will be able to help, although some are governed by military restrictions that do not allow potentially sensitive photographs of their country to be used by external bodies.

A laudable attempt to develop a World Index of Space Imagery (WISI) and a World Aerial Photography Index (WAPI) was initiated by the Food and Agriculture Organization (FAO) [499] in 1977. The results from the meetings that discuss this concept are published by the FAO as *Reports of an expert consultation on a World Index of Space Imagery (WISI), World Aerial Photography Index (WAPI) and thematic cartography for renewable natural resources development* [456]. The aim of this project is to ensure that aerial photography and satellite imagery archives and their holding organizations are recognized, thus enabling location of these resources for interested parties. An extensive WISI/WAPI conference network has been set up by the FAO, in co-operation with NOAA, for discussion and rationalization of these objectives. The reports from these expert consultations are very useful data sources and often contain maps (generally of African countries) detailing their imagery holdings. Useful appendices that list addresses are included, together with position papers that comment on progress with the WISI/WAPI objectives in various countries. The 1984 report proposed that an international conference would discuss the project in 1987, with the aim of bringing the WISI/WAPI proposal before the potential user community.

North America

The Earth Observation Satellite Company (EOSAT) [741] is the major producing, archival and distribution centre for all Landsat-related products in North America. EOSAT holds worldwide coverage of all Landsat 1, 2, 3, 4 and 5, MSS, RBV and TM data in its centralized archive at the EROS Data Center [742]. The Data Center is a USGS repository for remotely sensed aircraft and satellite data, and in addition to Landsat imagery, its holdings include worldwide coverage of Apollo/Gemini, Shuttle and Skylab material, together with all NASA, USGS and National High Altitude Photography (NHAP) of the United States. Skylab photography is detailed in the *Skylab Earth Resources Data Catalog* [453], a descriptive manual that provides a list of photography and furnishes additional information on the latitude, longitude, location, time and date of imaging. Skylab capabilities are explained at length and the sensor coverage maps that are provided may be linked, together with the indexed photographs, to an accompanying 16mm browse film on which the relevant photographs are stored. Although EOSAT is operating Landsats 4 and 5 through the EROS Data Center at the time of writing, it is due to transfer data processing and product generation to its permanent headquarters in Lanham, Maryland, before the launch of Landsat-6 (whose projected launch date is in the late 1980's).

There are several methods available to the potential user for locating a scene of interest and any subsequent order can be made either from the retrospective archive, or through the contemporary network. If a special request is made, imagery can be obtained over a particular area via the EOSAT Special Acquisitions service, details of which are available from the company. The most basic aid to searching is the Landsat Worldwide Reference System (WRS), a series of maps that chart the path/row number and scene centre of all Landsat MSS and TM overflights of the world. Potential customers are sent a WRS sheet of their area free of charge, and the scene centres of interest are chosen from this. The information is then used as the basis of a geosearch run on EOSAT's computerized database, which is returned to the customer as a printout list that provides an indication of what Landsat products are available for that area. A comprehensive non-image Landsat MicroCatalog microfiche reference system is also available on a subscription basis.

Designed primarily for use by those without access to EOSAT's database, the MicroCatalog can be used to determine worldwide Landsat data availability. An understanding of the WRS is required to use the MicroCatalog and a very useful descriptive manual, *Landsat MicroCatalog: a reference system* [449], was prepared by NOAA/NESDIS Landsat Customer Services [774] in 1983. An accompanying Microfilm browse file, covering the period 1972-1984, is available on 16mm film cartridges. This companion facility displays MSS band 5 or 2 data for a given scene, and as well as allowing the preview of geographic area coverage, it enables an assessment of cloud cover and data quality to be made. Each microfiche card contains approximately 1,200 scenes from MSS band 2 and TM bands 3 or 6, with band 6 being used for night-time imagery. EOSAT is also preparing a directory of 'value added' firms, which should be published shortly. EOSAT can furnish potential customers with Landsat products and services information, including order forms and price lists, upon request. Products do not have to be ordered direct from EOSAT's central headquarters, as it also has distribution contracts with centres in Argentina [512], Australia [515], Brazil [534], Canada [539], India [620], Indonesia [628], Japan [641] Pakistan [670], the People's Republic of China [579], the Republic of South Africa [684], Saudi Arabia [682], Thailand [700] and Europe (via the ESA Eurimage/Earthnet [524, 530, 589, 605, 609, 633, 640, 658, 669, 686, 692, 696, 703] user services).

It should be noted that all of the imagery supplied by EOSAT and its nominated distribution centres is covered by its Agreement for Purchase and Protection of Satellite Data. This agreement is in effect a copyright law that prevents the unauthorized dissemination of unenhanced Landsat data. The copying or transfer of original imagery is not allowed under the agreement, but products that have been enhanced are not covered. The agreement is not retroactive, and imagery products produced by the EROS Data Center prior to EOSAT's take-over of the Landsat programme on 27 September 1985 are not covered by the restriction and remain in the public domain. Copies of the Purchase Agreement may be obtained from EOSAT, and the legislation that provided the basis for protection of unenhanced data sets, the Land Remote-Sensing Commercialization Act of 1984: Public Law 98-365, is at present being reviewed by Congress and the Senate Committee on Science, Technology and Space.

NASA has funded a considerable number of spaceborne remote sensing experiments and while the imagery from these is not available on such a regular basis as Landsat data, it nevertheless forms several interesting data archives. A major NASA initiative was Shuttle Imaging Radar (SIR), and maps that show the coverage of SIR, together with the products generated by this experiment, are available from the National Space Science Data Center (NSSDC) [775] in Maryland. Over 10 million km^2 of the Earth's surface were covered at 1:500,000 scale by SIR-A and 6 million km^2 by SIR-B. The SIR-A experiment image swaths are available as film positives or negatives only, whilst SIR-B data may be obtained in both digital and optical formats. The SIR experiments were a progression from NASA's 1978 Seasat Synthetic Aperture Radar (SAR) satellite. This mission generated a tremendous amount of imagery, which is available as raw processed data, from which photographic and digital products have been generated. The public sale and distribution of Seasat SAR data are handled by the National Environmental Data and Information Service (NESDIS) [774] of NOAA. All of the imagery that is available has been processed at the Jet Propulsion Laboratory (JPL) [767], which produced the *Seasat SAR- Imagery Catalog* [447], a publication that provides full details of Seasat sensor characteristics, together with detailed information on the coverage obtained during the satellite's 106-day life. NASA also launched the Heat Capacity Mapping Mission (HCMM), a two-channel scanning radiometer, in 1978. This mission was retired in 1980, but still generated a large amount of data, which is available in digital, radiometrically corrected and radiometrically and geometrically corrected formats, together with photographic products that include quick-look prints, day visible/day infrared or night infrared passes, thermal inertia maps, temperature difference images and night infrared passes co-registered with a day image. U.S. users can obtain HCMM data from the NSSDC [775]; non-U.S. customers should apply to the World Data Center [804], or if in Europe, Eurimage/Earthnet. NASA's Scientific and Technical Information Branch [771] also produces the *Heat Capacity Mapping Mission (HCMM) anthology* [458] and *HCMM data users' handbook for applications - Explorer mission A* [451]; the former title provides full technical details of the mission and outlines the potential uses of the imagery. The Nimbus-7 Coastal Zone Color Scanner (CZCS) was yet another experimental satellite launched by NASA in 1978. The standard products available from the CZCS are 241mm black-and-white, positive and negative film transparencies that include all six bands. Each band covers an area of 700 x 1,363 km and contains approximately two minutes of data. Nine-track 1600-bpi CCT's that hold three two-minute scenes are also available, and all of this imagery may be obtained from NESDIS.

Apart from the SIR-A and -B experiments, NASA also funded the experimental Space Shuttle Large Format Camera (LFC) STS Mission 41-G in October of 1984. The imagery from this flight may be purchased from the Chicago Aerial Survey Inc. [752], which is licensed by NASA to sell LFC products. The film image scales run from 1:750,000 to about 1:1,200,000, but the photographs can be enlarged by up to 10 times without any significant loss in image quality. A useful accession aid (Product Code 99) that contains a microfiche of selected mission ephemeris data, geographic descriptors and footprint maps that show frame-by-frame locations of photographs may be obtained, together with tabulated lists of the complete range of products available from Chicago Aerial Survey Inc. In comparison with other types of imagery, LFC photography is relatively cheap as it is classified as experimental data; however, there are special conditions for data purchase and an agreement must be signed by prospective users that guarantees the distribution rights of the supplier under the Land Remote Sensing Commercialization Act of 1984: Public Law 98-315. Purchase agreement documents are supplied with each order form and these have to be signed before any data products are made available.

A complete list of the hand-held photography from several Shuttle missions is available from NASA as the *Catalog of Space Shuttle Earth Observation*

Hand-Held Photography. Space Transportation System (STS) 1, 2, 3 and 4 Missions [450].

The National High Altitude Photography (NHAP) programme is a multiagency Federal exercise which has been co-ordinated by the U.S. Geological Survey (USGS) [797] with the aim of providing complete conterminous photographic coverage of the United States. The programme, which was initiated in 1978 and is now complete, was conceived as a fund-saving replacement of previously existing agency programmes that often duplicated photographic coverage of different States. A steering committee oversees the programme and provides guidance, thus ensuring that the photographs are available to any interested user and that periodic updates of the database are made. The NHAP has a stringent set of product specifications, thus ensuring that data continuity is maintained for sucessive missions.

Black-and-white and colour infrared photographs are produced from the NHAP. Film transparencies or photographic reproductions of black-and-white or colour infrared photographs may be obtained in 9 x 9 inch formats or the enlargement

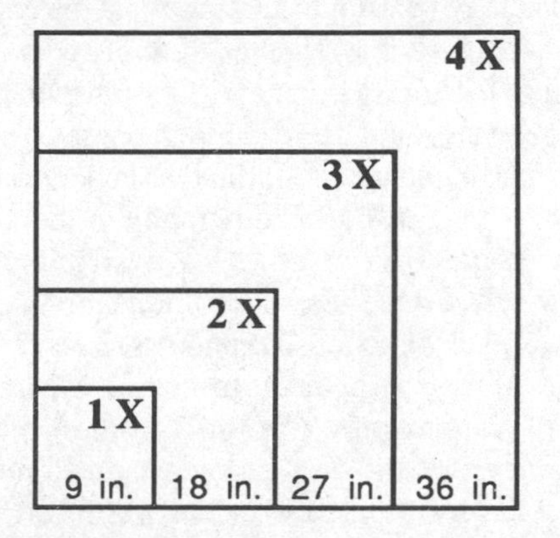

Figure 2 National High Altitude Programme photographic enlargement

sizes detailed in Figure 2. Special print sizes can be ordered at the customer's request. The accession aids that are available include a set of micrographic indexes to photography, together with mapline plots to 1:250,000 charts, which are in turn keyed to the 1:1,000,000 scale International Map of the World Series Index. The microfiche also contain the photographic identification information required to order prints, together with data on project year, roll and frame number. Microfiche accession products are available at a number of centres and orders for NHAP material should be directed either to the EROS Data Center [742] or the Aerial Photography Field Office [744].

A large photographic library is maintained by the USGS at Denver in Colorado, as part of the USGS Library System. The National Mapping Division of the USGS houses the National Cartographic Information Center (NCIC) [773], the primary information organization of the USGS. Cartographic and geographic information, including digital data, aerial photographs and maps, from USGS and other government, as well as industrial, sources is held at the NCIC. Most of the information can be rapidly retrieved by the use of the computerized database that has been established at the NCIC, a system that may also be accessed from many of the NCIC's cartographic information centres, which exist in forty-five States.

In addition to holding a microfiche guide to NHAP, the Technology Application Centre (TAC) [743] in New Mexico maintains an archive of over 36,000 aerial photographs and nearly 500 Landsat CCT's and images. More than 30,000 black-and-white aerial photographs of New Mexico, that vary in scale from 1:12,000 to 1:40,000 and which have been donated by the USGS and several other Federal agencies, are included in the archive, together with Soil Conservation Service (SCS) 15 and 7.5 minute controlled photographic mosaics. Over 6,000 high-altitude black-and-white, colour and colour infrared NASA aircraft photographs at a scale of 1:126,000 from U-2 and RB-57 reconnaissance planes are also held. The TAC continues to maintain its collection of second-generation master prints of worldwide hand-held photography from the Apollo, Gemini, Skylab, Apollo-Soyuz and Space Shuttle missions. Prints and transparencies of these are available in black-and-white, black-and-white infrared, colour and colour infrared, in sizes that range from 4 x 5 inches to 16 x 16 inches for prints and 4 x 5 inches to 8 x 8 inches for transparencies. Descriptive catalogues have been developed for these products and they do not contain photographs, but list geographical coverage, image quality and format, altitude, mission date, camera and film type information for the various photography. They are ideal guides and companions to the imagery and six volumes are available, one each for Gemini, Apollo and Apollo-Soyuz, and three separate editions for Skylab missions 2, 3 and 4 respectively. The TAC has an extensive photo-search capability as part of its affiliation to the National Cartographic Information Center (NCIC) [743], the central cartographic and photographic repository in the United States. TAC can offer computerized searches of the cartographic data stored in the EROS Data Center, which enables the location of Landsat MSS and RBV, together with Skylab, Apollo and Gemini photography for any area worldwide. A large archive of high- and low-altitude photography from a number of U.S. agencies can also be identified. The enquirer receives a printout from the computer search that details photograph scale, data quality, cloud cover, film type, number and date of exposure. A manual search service that utilizes microfiche index files to locate photographs of a particular area is also available. The cross-referenced fiche system contains NASA aircraft and USGS photographic coverage for New Mexico, including USGS aerial mapping photography indexes, together with a complete Landsat index covering the whole world and Skylab photography for the United States. Photographs can be ordered directly from the indexes, which cover cartographic information submitted to the NCIC database by government agencies and private organizations.

The library of the National Air and Space Museum (NASM) [751] houses the Planetary Image Facility of the Center for Earth and Planetary Studies. The intended use of this archive, which was established in 1983 under an agreement between NASA and NASM, is to provide a research report centre for scientists interested in planetary mission photography. Over 200,000 images and support data are available for consultation at the NASM and its regional centres in the University of Arizona [781], Brown University [778], Cornell University [779], the University of Hawaii [782], the Jet Propulsion Laboratory [767], the Lunar and Planetary Institute [780], USGS [783] and Washington University [784]. The materials available include selected prints and transparencies from Ranger, Surveyor, Lunar Orbiter and Apollo Missions; 8 x 10 inch prints from Mariner 4, 6, 7, 9 and 10; Venus from the Viking Landers/Orbiters; Voyager 1 and 2 mission photography of Jupiter, Saturn and their satellites; together with computerized topographic data of the Pioneer missions to Venus. Unfortunately reproduction facilities are not provided, but the materials can be ordered from the NSSDC [775], with the assistance of a qualified on-site photolibrarian, who may also provide assistance with online computer searches for imagery and advice on the use of the browse file and microfiche support aids.

Significant quantities of aerial photographs are also available in several of the larger U.S. public and university libraries. Notable collections exist at the University of Georgia Library (190,000 aerial photographs) [794], Virginia State University (33,000 mainly geological aerial photographs) [801], the University of Wisconsin (100,000 Landsat images and 80,000 aerial photographs) [795] and the Bernice P. Bishop Museum (70,000 aerial photographs) [750].

A major resource in Canada is the National Air Photo Library (NAPL) [564]. Over 4.3 million aerial photographs of Canada, dating back to the 1920's, are stored in this archive. Black-and-white photography is available for the whole of Canada; colour photography may be obtained for some areas and colour infrared imagery covers even more selected regions of the country. The photographs are at a variety of scales and enlargements, and mosaics and index maps may be obtained. A series of accession aids are available, including catalogues, microfilmed index maps and microfilmed imagery. The general-coverage system catalogues are divided into ten provincial packages, together with an Airborne Remote Sensing Photography Catalogue that covers the whole of Canada. Each catalogue displays all of the coverage since 1966, as well as selective examples obtained prior to 1966. Colour-positive reproductions of index maps are included on the lines and photo centres of the NAPL archived imagery. The accompanying 16mm Cartridge of Microfilm and Imagery features aerial photographs listed in ascending order on the cartridge and roll number catalogue. A very useful descriptive booklet, *General Coverage Systems Catalogue (Catalogue de la Couverture Photographique Aérienne du Canada)* [445], that gives fuller details of these aids, together with photograph ordering procedures, is published by the Surveys and Mapping Branch of the NAPL. Requests for photographs are usually processed within four weeks and information on prices, products and Federal photographic coverage may be obtained from the NAPL, or its reference centres in Manitoba [542], British Columbia [546, 573], Nova Scotia [545], Alberta [541, 544], North West Territories [543], Ontario [540], Yukon [557] and Newfoundland [554].

Several Canadian universities also house extensive collections of aerial photographs for teaching purposes and as provincial archives. Over 1 million aerial photographs are stored in centres at the Université du Québec (400,000) [571], Dalhousie University (12,000) [552], the University of Ottawa (245,000) [574], the Université Laval (127,000) [572], and the University of Toronto (227,000) [575].

Western Europe

There are several major sources of aerial photography and satellite imagery within the U.K. and one of the most important of these is the collection held by the Ordnance Survey (OS) [731] as its Central Register of Aerial Photography for England (CRAPE). This archive includes complete coverage of England at scales from 1:3,000 to 1:30,000, panchromatic and infrared photography of the Commonwealth developing countries and selected Landsat 1, 2 and 3 photographic products of developing countries. The Overseas Directorate of the Ordnance Survey, formerly the Directorate of Overseas Surveys (DOS), merged with the OS in 1984. This organization is primarily concerned with surveying and mapping in developing countries and produces some 600 maps annually, mostly from aerial photographs. The Ordnance Survey of Northern Ireland (OSNI) [730] maintains the Northern Ireland Register of Aerial Photography. This database is interesting as it contains several Landsat TM and MSS scenes, AVHRR data and SPOT simulation imagery for selected areas of Northern Ireland, together with a large set of panchromatic vertical photography at scales from 1:10,000 to 1:20,000; several areas are also available as oblique prints at a variety of scales.

The archival storage and development of a national collection of vertical aerial photographs including Royal Air Force (RAF), Ordnance Survey (OS), archaeological, historical, architectural, cartographic and landscape images is performed by the Air Photographs Unit of the Royal Commission of the Historical Monuments of England (RCHME) [706]. Around 500,000 oblique panchromatic, colour and colour infrared prints and negatives dating back to 1924 are held at Fortress House, while approximately 2.5 million prints and negatives of vertical panchromatic photographs are stored at Acton [707]. The latter materials are mainly from the 1946-1965 period, including post-war RAF surveys, but some prints are current to 1985. The RCHME adds between 20,000 and 30,000 items to its photographic database each year and carries out its own research programme, which at the time of writing is engaged upon the development of a map-based digital index of its holdings.

The ADAS Aerial Photography Unit [705] and Resource Planning Group [735] of the Ministry of Agriculture, Fisheries and Food (MAFF) hold substantial amounts of both airborne and spaceborne remotely sensed data in their respective archives that are, or have been, used for a variety of government projects. Over 70,000 vertical panchromatic and colour aerial photographs are held by the Central Register of Air Photography for Wales [717]; flight index maps are available for the imagery, which has been flown since 1945. The Central Register is also able to supply all RAF photography and vintage OS imagery for the period 1954-1969. A similar archive, the Central Register of Air Photography of Scotland, is held by the Air Photography Unit [708] of the Scottish Development Department (SDD). The Department of Geography [739] at the University of Keele holds 5.5 million RAF photographs of wartime Europe, together with airborne Synthetic Aperture Radar (SAR), SPOT simulation and thermal infrared linescanner images and CCT's. Its satellite archive consists of selected Seasat and Landsat MSS, RBV and TM images and CCT's. The archived imagery, together with worldwide image indexes, may be consulted using microfiche image catalogues for Seasat, SIR-A, Large Format Camera (LFC) and the Metric Camera.

The National Remote Sensing Centre (NRSC) [703] at Farnborough maintains an extensive archive of satellite data in its capacity as a National Point of Contact (NPOC) to Eurimage [721]; it has access to imagery received within the Eurimage coverage region, which is described later in this section. The data holdings include over 1,100 MSS and RBV images from Landsats 1, 2 and 3, together with more than 200 MSS and TM images from Landsats 4 and 5. The NRSC also holds a two-week rolling archive of weather satellite data and has recently become a SPOT data distribution centre which will maintain a U.K. archive of SPOT imagery. A variety of browse facilities are available at the Centre, including quick-look archives of all U.K. Landsat images acquired through Earthnet since 1975, microfiche catalogues of the MOMS, Seasat and Metric Camera missions, access to the ESA-IRS [498] and LEDA [317] computerized imagery index and small hard-copy prints of all NRSC-archived scenes. An excellent publication, the *NRSC Data Users' Guide* [455], that fully describes the activities and services of the Centre has been prepared at the NRSC. The *NRSC Data Users' Guide* is a model example of how to communicate the maximum amount of information in a brief and readily understood form. It starts with background details concerning the NRSC, including its terms of reference, policies and an outline of NRSC working groups and the Centre's role in what was the Earthnet network. The second section is an annotated list of current and past major satellite systems, together with their respective technical specifications and details of the products generated from them. The third part of the *NRSC Data Users' Guide* is in the form of a series of appendices that present a set of fifty-four maps that illustrate the whole world and show the path/row coverage of SPOT and Landsats 1, 2, 3, 4 and 5 over Europe, Africa, the Middle East, Asia, Australasia and the Americas. In addition to this, the appendices list the NRSC imagery archive and detail tape formats, photographic specifications and prices. The sections are all suitable for ring binding and this option ensures that the guide can be kept current by removing out-of-date sheets and replacing them with updated information. The catalogue of Landsat imagery held at the NRSC is also available at the British Library's Map Room [714], together with the EROS Data Center MicroCatalog of all available Landsat imagery from 1972, its accompanying 16mm browse film, the *Skylab Earth Resources Data Catalog* [453], its companion 16mm browse film and the *Seasat SAR Imagery Catalog* [447]. All of these resources, together with a series of colour composite images, are available on open access in the map room.

The structure and function of the ESA's former Earthnet network, a system responsible for the acquisition, pre-processing, archiving and dissemination of satellite data to a series of National Points of Contact (NPOC), has already been described in detail in Chapter 2, together with details of the new structure adopted in 1987 following private-sector involvement. The central archive and processing point for satellite data within Earthnet is the Earthnet Programme Office (EPO) [498] at Frascati in Italy. However, since the recent commercialization of this service, by Eurimage [721], the Eurimage NPOC (the former Earthnet NPOC) should be contacted for ordering data. The EPO will be maintained, but many of its services will be provided by Eurimage staff operating from within the EPO.

The data collection at Frascati includes over 300,000 Landsat images, 1,000 Seasat swaths, 3,000 Nimbus CZCS images, 1,540 HCMM passes and 1,000 Metric camera photographs. All of these data are available to end users through the Eurimage NPOC in Austria [524], Belgium [530], Denmark [589], France [605], Germany [609], Ireland [633], Italy [640], The Netherlands [658], Norway [669], Spain [686], Sweden [692], Switzerland [696], and the United Kingdom [703], as well as through the central headquarters of Eurimage. Five ground receiving stations handle incoming data, and three of these, Fucino, Kiruna, and Maspalomas, respectively in Italy, Sweden and the Canary Islands, gather Landsat data. Fucino receives data from Central and Eastern Europe, the U.S.S.R., North Africa and the Middle East; the photoprocessing of these data is performed by the Italian NPOC. While Fucino receives both MSS and TM data, Maspalomas gathers only MSS data of North and West Africa. Both stations transmit their data directly to the Eurimage/Earthnet distribution network at ESRIN-EPO. The U.S.S.R., Northern Europe, the Arctic, Iceland, Greenland and Scandinavia are covered by the Kiruna station, and acquisitions at this centre are sent directly to the particular NPOC requiring the data, although the requests for imagery should be addressed to Eurimage/EPO. Permanent browse facilities are located at Frascati, and quick-look files may be consulted, together with the frequently updated EROS Data Center MicroCatalog. The centre also houses the LEDA [317] online catalogue of Eurimage/Earthnet data availability as part of the ESA-IRS [498]. This database contains over 800,000 items and is a computer-readable list of all MSS, RBV and TM imagery from Landsats 1, 2, 3, 4 and 5. The file may be searched for geographical location information via the conventional World Reference System (WRS) of path/row numbers, or by latitude/longitude coordinates. Printouts include data on the satellite and sensor used, cloud coverage, image quality, sun azimuth elevation at the time of imaging and the date of imaging.

A branch of the ESA, the German Eurimage NPOC, DFVLR [609], has been responsible for two Shuttle-borne remote sensing experiments, the Metric camera and the Modular Opto-electronic Multispectral Scanner (MOMS). During the Metric camera flight over 1,000 colour infrared and panchromatic images were obtained, covering an area of 11 million km^2 over North, Central and South America, Europe, Africa and the Middle East. Metric camera photographs in standard 23 x 23 cm (1:820,000 scale) transparency or print formats may be ordered through the relevant NPOC, or directly from the DFVLR, together with a descriptive catalogue and set of microfiche containing the acquired imagery. A stereoscopic capability is available with the images, as a 60% overlap in the flight direction was obtained during the mission. MOMS was flown in the Shuttle in both 1983 and 1984, and the two missions generated imagery covering an area of over 3.5 million km^2 between them. Data products are available from the DFVLR and include film negatives and black-and-white prints at 1:800,000 scale, together with enlargements of these, and CCT's at 1,600 and 6,250 bpi. Eurimage/Earthnet also distribute Seasat SAR, HCMM, and Nimbus CZCS products, in a variety of data formats. Seasat SAR data were received at the Oakehanger receiving station and an archive of fifty-three swaths, or 256 minutes of data, exists. A number of accession aids which greatly facilitate any search for images are available for Seasat data from Earthnet User Services in Frascati. One of the most useful is a microfiche catalogue that contains 505 Seasat SAR images of the ESA/Earthnet coverage area. Fourteen fiche are produced; the first is a computer listing of orbital parameters, frame centre co-ordinates and archive numbers, while fiche 2-13 show the 505 40 x 60 km areas and provide information on orbit number and acquisition date.

Lannion in France gathered over 1,500 HCMM passes over Western Europe and Northern Africa and advice on the availability of these data and all of the NASA experimental satellite imagery can be obtained either directly through Frascati or via the respective NPOC. A useful summary of Earthnet's structure and activities is provided by *Earthnet - the story of images* [444], and although this booklet is somewhat dated with regard to the latest developments in remote sensing, it nevertheless presents a detailed account of Earthnet's terms of reference, together with information on the receiving and processing of a variety of types of imagery from a number of satellites and sensors, an excellent description of LEDA and a list of products available from the EPO.

A wide range of high-resolution panchromatic and multispectral satellite data products has become available from SPOT Image [598] following the successful launch of SPOT 1 on 22 February 1986. Data are received from the satellite at two principal stations, located at ESRANGE [693] near Kiruna and Aussaguel near Toulouse. Each of these stations can acquire up to 700 scenes each day, which may be archived and placed on the SPOT catalogue if they meet certain quality requirements. SPOT is currently establishing a network of international direct receiving stations that will archive images of their respective coverage area, act as SPOT Image distributors in their country and supply SPOT Image with images for distribution, together with the details required to update the SPOT Image catalogue. At the time of writing receiving stations have been agreed for Canada [539], China [583] and India [620], while negotiations are under way with Brazil [532], Japan [641], Pakistan [670], Saudi Arabia [682] and Australia [520]. These stations should also distribute SPOT data, together with a well established network of nearly forty other distribution centres in Africa [650, 667, 684, 702], North America [556, 789], South America [512, 536, 531, 578, 654, 672, 813], Asia [641, 649, 657, 673, 699, 700], Europe [525, 528, 590, 596, 609, 618, 631, 640, 658, 668, 675, 679, 687, 694, 696, 703, 729, 815] and the Middle East [592, 635].

The SPOT panchromatic and multispectral scenes are preprocessed to four basic levels:

Level 1 A:	A normalized raw data level that has no geometric correction and is intended for basic radiometric studies and stereoplotting
Level 1 B:	Basic-level data suitable for photointerpretation that have been corrected radiometrically and geometrically
Level 2:	A step further than Level 1 B, in which corrections are applied to cartographic standards to produce a precision-processed image map
Level S:	A corrected image intended for multidate use in combination with a SPOT reference scene

In addition to the high-quality data described above, SPOT plans to offer certain special and derivative products such as stereo-pairs, composites and image mosaics, that will illustrate the unique qualities of the data. These data are available in a range of media, including CCT's at 1,600 and 6,250 bpi, together with a number of photographic products. Prices for these are available from the respective distributor and are also quoted on the 'SPOT Image Product Order Form', which may be obtained from SPOT Image and its representatives.

A number of accession aids are available for SPOT data and the most basic of these is the SPOT Grid Reference System (GRS), a series of fourteen column and line reference maps that cover the whole world. A much more advanced online aid, the SPOT Image Catalogue, contains a worldwide listing of scenes acquired and archived by SPOT Image and its nominated receiving stations. The catalogue may be searched by SPOT if the enquirer completes an 'Inquiry Form for SPOT Data Search', or the user can be connected directly to the catalogue and may apply for a subscription to the Browse and Mail System (BRAMS) Computer Service, by completing a 'BRAMS computerized service subscription application form'. The fully computerized catalogue is operational twenty-four hours a day and 365 days a year; it may be searched via the International Telephone Network (ITN), the International Public Switched Telephone Network (PSTN), or a Packet Switching System (PSS) connected to the French Transpac network, or other networks such as Tymnet, Telenet, Euronet and Datapac. Various types of access are available, consisting of two levels of assisted mode and one direct mode. The assisted-level modes are menu-driven packages that are suitable for users unfamiliar with the system, whilst the direct mode uses a more basic and rapid system vocabulary. Frequent SPOT Image customers may also subscribe to an 'electronic mailbox' (EMB) facility, which allows messages to be passed between SPOT and the user. Hard-copy quick-look prints of images from the catalogue may be purchased from SPOT in packs of five scenes, and a U-MATIC video-cassette which offers 500 quick looks over a specified geographical area is also currently available. A special service that images particular areas on request is available for those users who cannot find the imagery they require in the SPOT Catalogue. Imagery that has to be acquired in this manner is ordered via a 'SPOT 1 Programming Request' (PR) form. The potential users are required to define their geographical area of interest, preferably delineating it on a map, and can also state the maximum cloud cover they require, for how long they wish an area to be surveyed and the surveying method they wish to be used. Once PR conditions are accepted by SPOT Image and the data are obtained, the user is committed to purchase the scenes that have been collected.

Like EOSAT, SPOT Image data products are covered by copyright laws which limit the usage, copying and dissemination of the data. Copies of their general terms of sale are reproduced on most of the forms used for ordering products, which are all available from SPOT Image and its nominated product distributors.

The Institut Géographique National (IGN) [597] in France has an aerial photographic collection that exceeds 3 million prints and negatives of national territory and some countries that were formerly under French administration. Digitization of this archive began in 1984 as part of the production of a French Satellite Imagery Catalogue. According to a report by Guyot (1984), aerial photographic information can be retrieved by the mission name, scale, administrative unit and name of the corresponding 1:50,000 scale map via a computer terminal.

Dominantly educational institutions such as the ITC [500] and JRC [502] hold data for teaching purposes; these are often mainly aerial photographs, but the JRC also has a diverse archive of Landsat MSS and TM, NOAA thermal MSS, SPOT simulation and SAR-580 images of the EEC and Sahel. Many of the other European groups and organizations that hold considerable amounts of aerial photographs and satellite data are listed in Part III.

Eastern Europe and the U.S.S.R.

Sources of remotely sensed imagery in Eastern-bloc countries are very poorly advertised owing to the politically sensitive nature of such imagery. However, a substantial archive of 800 Landsat 1, 2, 3, 4 and 5 and Cosmos scenes, together with over 25,000 An-30 MKF-6 and An-2 non-metric aerial photographs, is maintained at the Földmérési Intézet (FÖMI) Remote Sensing Center [618] in Hungary. These are publicly accessible collections, with the exception of the Cosmos imagery and Metric aerial photographs, which are classified.

Quite recently, a Soviet organization, Sojuzkarta [810], has begun to market what were previously unobtainable examples of Soviet space photography. The data are apparently of a very similar resolution to Metric Camera photographs, but it seems that there is fairly extensive coverage of many countries. The sensors used include the K-140 (60 m resolution), the three-channel KATE-200 (30 m resolution) and the six-channel MKF-6 (20 m resolution); a price list for these products as contacts, prints and reprints is also available. Unlike many other forms of remotely sensed data these products are free from copyright and therefore carry no commercial restrictions.

Africa

Obtaining information on the holdings of remotely sensed data in African countries is very difficult. One of the most useful guides to archives is the *Reports of an expert consultation on a World Index of Space Imagery (WISI), World Aerial Photography Index (WAPI) and thematic cartography for renewable natural resources development* [456], mentioned towards the start of this chapter. A useful annex in this publication contains a list of aerial photographic coverage for Sénégal [683], Mauritius [651], Mali [652], Guinea [616] and the Gambia [608]. Addresses of the holders of the aerial photography are listed, together with maps that indicate the scale, type and coverage of the aerial photography for each country.

The Regional Centre for Services in Surveying, Mapping and Remote Sensing [645] in Kenya has a considerable collection of remotely sensed data used for its various services to central and southern African countries. The imagery collection is largely Landsat data and includes:

1. A worldwide collection of multi-band 70mm film positives of all Landsat 1 and 2 images acquired between 1972 and 1977
2. All MSS Band 5, 70mm negatives with a cloud cover of less than 30% for the Eastern, Central and Southern African region, for the period 1972-1982
3. All the RBV 70mm negatives acquired by Landsat 3 with a cloud cover of 30% or less
4. A selection of high-quality standard FCC and transparencies for the above catchment region
5. A variety of special products, including enhanced transparencies, film mosaics and different colour renditions, that have been composited in an in-house photographic laboratory

Another large collection of over 1,000 Landsat images, together with several hundred aerial photographs and image mosaics of western Africa and Burkina Faso, is held by the Centre Régional de Télédétection, Remote Sensing Center of Ouagadougou [538].

The Institut National de Geodésie et Cartographie [647] in Madagascar is primarily concerned with taking aerial photographs and using them for the production of thematic maps. Over 60,000 panchromatic aerial photographs at a variety of scales are used in the map-making process and more than 600 image mosaics, which represent combinations of these photographs, also exist. A small part of south-west Madagascar is covered by Landsat MSS data and an archive of 4,000 historical aerial photographs is available at a variety of scales.

The Satellite Remote Sensing Centre (SRSC) [684] is operated by the National Institute for Telecommunications Research for the Council for Scientific and Industrial Research (CSIR) in the Republic of South Africa. The Centre is capable of receiving Landsat MSS images of Southern Africa and currently has an archive of over 38,000 Landsat 1, 2, 3, 4 and 5 MSS scenes stored on HDDT. Imagery generated from the HDDT's is available in a number of standard black-and-white and false-colour composite formats at scales between 1:250,000 and 1:1,000,000. A catalogue that lists MSS imagery acquired and stored at the centre may be purchased, or users may request free extracts from the catalogue that cover only their area, or areas, of interest. The SRSC offers customized products and an accompanying expert research capability in addition to the standard products and services. The SRSC is also a distributor for SPOT imagery within Africa, although other SPOT distribution centres are located in Malawi [650], Nigeria [667] and Tunisia [702].

South America

The Departamento de Aplicações de Dudos de Satélite of the Instituto de Pesquisas Espaciais (INPE) [532] has been receiving imagery from Landsat satellites since 1973. Both MSS and TM data are recorded at Cuibá, Mato Grosso State, which allows full coverage of Brazil and a substantial amount of South America as a whole. INPE's databank consists of approximately 200,000 images which may be accessed via computerized search facilities. The research and development programmes at INPE are supported by an airborne data acquisition system consisting of a light plane carrying one panoramic camera, two metric cameras, four 70mm cameras, one multispectral camera, one thermal imaging scanner and a thermal radiometer. Data from all of these sensors are housed at INPE, and a great deal of colour infrared and black-and-white aerial photography is also available. In addition to receiving Landsat images, INPE is negotiating with SPOT Image for the reception and distribution of SPOT data. Another organization that distributes SPOT in Brazil is Sensora [536], based in Rio de Janeiro. A diverse collection of 4,150 aerial photographs, twenty-two radar swaths and fifty-two satellite images is held by the Departamento de Desenvolvimento de Systemas of the Instituto de Desenvolvimento de Pernambuco (CONDEPE) [535].

The Instituto Geográfico Militar, Sección Sensores Remotos [576] in Chile has a useful archive of remotely sensed data that is publicly accessible. Nearly sixty Landsat scenes cover ninety percent of the country, which is totally covered by various scales of black-and-white photography and more selectively by colour infrared photography. A very large collection of 66,000 aerial photographs is also available in the archival division of the Instituto Nacional de Investigación de Recursos Naturales [577]. The Servicio Aerofotogramétrico de la Fuerza Aerea (SAF) [578] in Santiago is the nominated vendor of products from SPOT Image in Chile.

Asia and the East Pacific

The National Remote Sensing Agency (NRSA) [620] receives both MSS and TM data from Landsat and acts as EOSAT's international distribution centre in India. The NRSA also represents the interests of SPOT Image in India and the organization receives, archives and distributes SPOT imagery. The extensive Landsat 2, 3, 4 and 5 archive of over 4,000 HDDT's covers all of India, as well as parts of Bangladesh, Pakistan, Nepal and China, while SPOT imagery is available for a slightly larger area. A computerized geographic search and enquiry facility is operated at the NRSA, enabling the potential customer to search the archive by path/row, area rectangle or map enquiry. Browse facilities and data catalogues of Landsat coverage are available for consultation, together with quick-look films of all scenes acquired by the NRSA. In addition to Landsat and SPOT data, the NRSA has a collection of 110 aerial scanner CCT's/HDDT's, 200,000 panchromatic colour and colour infrared aerial photographic negatives and nearly 100 satellite and aerial mosaics at various scales. The major agency for the archiving and dissemination of aerial photographs in this country is the Survey of India (SOI) [625], which also allows the NRSA and other organizations to acquire imagery. Nearly all of India is covered by high-altitude aerial photography at a scale of 1:60,000, and an index of this coverage is maintained by the SOI. Unfortunately the public availability of much of this aerial photography is questionable, owing to governmental and military restrictions.

The Remote Sensing Applications Division of the Pakistan Space and Upper Atmosphere Research Commission (SUPARCO) [670] has a large archive of satellite data, including over 2,200 Landsat MSS scenes and HDDT's, fifty-one SIR-A swaths and several Metric camera and TIROS-N images. The Landsat archive provides repetitive coverage of Pakistan from 1972 onwards and is backed by a well equipped laboratory for photointerpretation and digital image processing. A further collection of Landsat 2, 3, 4 and 5 data is held at the Thailand Remote Sensing Centre (TRSC) [700] in Bangkok. Over 34,000 Landsat scenes covering areas of Nepal, Sri Lanka, Tibet, China, Burma, Bangladesh, Indo-China, the Philippines, Taiwan, Indonesia, Malaysia and Thailand are stored at the TRSC, which became a distributor of SPOT products in 1986.

Australia and New Zealand

The Centre for Remote Sensing [514] at the University of New South Wales in Australia holds over 100 CCT's that provide 1:500,000 scale Landsat coverage of eastern Australia, together with considerable amounts of aerial photography, SIR-A and SIR-B data and aircraft scanner imagery. A small collection of 5,000 aerial photographs is held by the Map Library [519] in Sydney, and a more substantial collection of 28,000 aerial photographs is archived at the University of Auckland's Geology Department [662] in New Zealand.

6.2 Global Meteorological Satellite Data

Meteorological imagery is currently available from two main satellite groups: geostationary meteorological satellites and near-polar-orbiting satellites. A detailed account of the satellites that supply the data, together with their respective areas of coverage, is provided in Chapter 1. The two different types of satellite are complementary in their data provision, as near-polar orbiters produce imagery for polar regions and higher latitudes, while geostationary satellites are used for the acquisition of equatorial and temperate zone imagery.

The National Environmental Satellite Data and Information Service (NESDIS) [774] receives data from the polar-orbiters and geostationary TIROS-N (NOAA 6 and 7), Advanced TIROS-N (NOAA 8 and 9) and GOES satellite series. An archive of over 8 million negatives and 25,000 CCT's is maintained at NESDIS. Two Meteosats have been launched by the ESA since 1977; Meteosat-1 is now retired, but a complete collection of Meteosat data, both digital and photographic, from 1977 onwards is maintained at the European Space Operations Centre (ESOC) [496]. The various meteorological offices and departments in countries all over the world will generally maintain an archive of meteorological satellite data related to their area. To list these stations and the data products available from them, which range from full-disc images, grid enlargements, mosaics, film loops and sectorized enlargements, would almost fill this book, and as many of the applications of this sort of data are real-time, it is more useful to describe the available data relay systems than the retrospective archives.

Images from meteorological satellites are generally relayed to the ground in both digital and analogue forms. Analogue transmissions are generally less detailed and precise than digital data, but they do not require the complex processing and reception equipment of the digital images. This means that reception of analogue data is cheaper and easier than digital transmissions and display of the images is possible on any suitable facsimile recorder.

Polar Orbiting Meteorological Satellites

The NOAA TIROS-N series satellites are used to provide visible and infrared imagery, together with atmospheric sounding data and meteorological satellite data relay and collection facilities. Imagery is provided by a four- to five-band scanning radiometer, the Advanced Very High Resolution Radiometer (AVHRR), which is sensitive to visible and infrared radiation. AVHRR data is time-multiplexed on board the spacecraft for digital broadcast as part of the High Resolution Picture Transmission (HRPT) at 360 lines per minute, providing a resolution of 1.1 km at nadir, in four or five spectral bands. Receiving stations may also acquire analogue data at 120 lines per minute, 4 km resolution, via Automatic Picture Transmission (APT), or by Weather Facsimile (WEFAX) broadcasts, which are described later in this section. The atmospheric sounding data, which are required for weather forecasting, can be obtained as either split-phase or linearly-polarized beacon information via Direct Sounding Transmissions (DST). An excellent publication, the *Polar Orbiter Newsletter* [380], has recently been introduced by NESDIS. This bulletin comments on polar orbiters' image quality, and also lists what imagery has been recorded and archived in the two-week period between each successive issue.

The Soviet Meteor satellite series transmit APT data at irregular intervals and form the basis of the U.S.S.R.'s operational satellite service. Visible and infrared cloud images are provided by the series, together with snow and ice cover pictures, cloud temperature, cloud height and solar radiation data. Unfortunately there is very little information on the broadcasts from these satellites, and apart from certain areas of the U.S.A., the Meteor transmissions are regarded as somewhat erratic; however, they are still processed and distributed over the Global Telecommunications System (GTS) of the World Meteorological Organization's (WMO) [503] World Weather Watch (WWW) programme. Data from the TIROS-N series are also transmitted via the GTS as part of the voluntary data exchange. This involves co-operation with France and the United Kingdom, as data are initially received at the NOAA satellite processing centre in Washington D.C. and are then transmitted to the ARGOS centre in Toulouse. The ARGOS centre is a satellite data collection and platform location system which processes the data that are later received by users, as well as the information transmitted over the GTS.

Geostationary Meteorological Satellites

Only one of the U.S. series of GOES satellites, GOES-6, is still providing imagery. However, GOES 2, 3, 4 and 5 are used as Data Collection Platforms (DCP's), which collect and relay meteorological data. Every half hour the sensors measure cloud cover and earth cloud radiance for the Western hemisphere, as well as providing temperature and water content information. GOES-6 has an imaging capability which transmits both high- and low-resolution WEFAX images, while GOES 3 and 5 also provide low-resolution, 14 km pixel size, image services. The broadcasts from the GOES series comprise five main analogue products which include: GOES VISSR imagery (visible and infrared); TIROS-N polar orbiter imagery; NMC forecast and analysis charts; TBUS messages that give TIROS-N orbital elements; and operational messages. In addition to providing analogue WEFAX imagery, the VISSR data are transmitted in a higher-resolution stretched bit rate digital form commonly known as Stretched VISSR (S/VISSR) data. A service very similar to the U.S. GOES series is provided by the Japanese Geostationary Meteorological Satellite (GMS). Both high- and low-resolution visible and infrared imagery are broadcast by GMS-3, which provides DCP services in conjunction with GMS-2 and the GOES series.

The ESA geostationary satellite Meteosat-2 provides visible and infrared full Earth disc images every half hour. The channels available on the satellite include two identical visible bands and thermal infrared and infrared water vapour channels that can be used in place of one of the visible channels. The infrared images provide a 5 km resolution at the subsatellite point, while the visible image has a resolution of 2.5 km (provided the infrared water vapour channel is not in use). Data are transmitted to the ESA ground station in Darmstadt and are re-broadcast after coastlines have been added and the image has been divided into different sectors. Data are relayed by GOES-4, which was moved to act as a DCP for data, following a decay in the orbit of the previous DCP, Meteosat-1, and faults with the DCP facility on Meteosat-2. Two telecommunications dissemination channels broadcast the data to receiving stations in high-resolution digital form to Primary

Data Users Stations (PDUS) and lower-quality analogue WEFAX form for reception by Secondary Data Users Stations (SDUS).

6.3 Guide Books and Catalogues of Remotely Sensed Data

Several useful guides to available aerial photography and mapping have been produced during the last decade. The most up-to-date of these is a new guide book, the *Guide to obtaining USGS information* [443], which has been compiled by K. Dodd, H. K. Fuller and P. F. Clarke, and published as *USGS Circular 900*. Amongst the products listed are aerial photographs, satellite images, catalogues, indexes and digital data. A great number of addresses of the recently reorganized USGS distribution points are included, together with information on remotely sensed data archives in the USGS Library System and at the NCIC [773].

Pilot Rock Inc. published *Everyone's space handbook: a photo imagery source manual* [448] in 1976. This somewhat dated guide identifies a number of the agencies and organizations that act as repositories and distributors of imagery from space, and includes a brief introductory history of orbital remote sensing and aerial photography. The *High Schools Geography Project: sources of information and materials; maps and aerial photographs* [446] is a similar, but slightly older publication that is inclined toward aerial photographs. This guide, which was published by the Committee on Maps and Aerial Photographs of the Association of American Geographers in 1970, is an annotated bibliography of aerial photography, remote sensing, maps and mapping. Some information concerning sources of aerial photography within the U.S. is provided, but many of these are now out of date.

Reference

Guyot, L. Information on some aspects of the future SPOT Imagery Catalogue, the ongoing digitization of the IGN Aerial Photographic Library, the existing Databank on Natural Resources at BRGM., and on the thematic evaluation of a SPOT simulation campaign. *Reports of an expert consultation on a World Index of Space Imagery (WISI), World Aerial Photography Index (WAPI) and thematic cartography for renewable natural resources development*. RSC series **28**: 92-97

7 Audio-Visual Materials

The requirement for pictorial material is very great in a discipline that is as essentially visual as remote sensing, and there are a wealth of illustrative resources including slide packs, filmstrips, hard-copy products, videos, films, maps, mosaics, picture atlases and training manuals. These have been produced by both large and small organizations and groups. Most of the available materials are primarily useful as educational or introductory aids to remote sensing. Some are more general in outlook, but others concentrate on specific aspects of the discipline. Several of the more ambitious and wide-ranging publications can be used in combination with other materials to develop interpretative and other skills. Although many materials are readily available as a series from organizations such as the U.S. Geological Survey, the products of smaller groups may only be issued on an irregular basis and are generally less easy to locate. The problems associated with discovering the sources of audio-visual materials are exacerbated by the fact that they are subject to very poor bibliographic control. To an extent the problem has been eased by the advent of online information retrieval, but regular correspondence with issuing organizations, allied with the frequent scanning of publicity materials, is required in order to keep fully up to date with what is available.

7.1 Slides and Filmstrips

One of the most useful and readily available sources of slide sets is the recently established RESORS [320] database (which is more fully described in Chapter 4) run by Gregory Geoscience for the Canada Center for Remote Sensing (CCRS) [539]. The database currently holds over 6,000 35mm slides pertaining to all aspects of remote sensing, and a printed list of the slide sets held by RESORS is available from the CCRS on request. Regular updates are held, and in the period 1985-1986, 899 slides were added to the system. There are over 150 slide sets on the database; thirty-two of these are commercially produced and the remainder are either derived from CCRS projects, or provide illustrative material for publications, presentations and lectures. The slide database, RESORS-S 35mm slides, is searched using a weighted keyword facility, and the information that may be retrieved concerning slides includes title, description, source/donor, accession number and category code, processing/publication date and quality. Slide sets may be borrowed for a month and can be ordered online via the seven-digit RESORS accession number. At the time of writing an additional copying service is available, whereby if the user forwards duplicating film to RESORS, slides may be copied and the undeveloped film returned to the user for no charge.

The CCRS also contributes to a very comprehensive archive of over 2,800 35mm slides of colour infrared Landsat imagery that covers the entire landmass of Canada, with the exception of the northern part of Ellesmere Island, which is beyond the range of the Landsat satellite. The slides are available from the National Air Photo Library (NAPL) [564], together with copies of selected cloud-free and low-snow-cover images chosen from the thousands of Landsat scenes archived at the CCRS. A descriptive leaflet that illustrates the coverage of the slides is available on request and it also provides a price list for the eleven regional sets, together with brief details of other products that are marketed by the NAPL. Three further sets that cover several regions each can be bought, and the numbers of slides each contains, together with regional pack sizes, is given below:

Newfoundland *	53 slides
Maritimes *	19 slides
Quebec *	135 slides
Ontario *	86 slides
Manitoba *	58 slides
Saskatchewan *	55 slides
Alberta*	57 slides
British Colombia *	90 slides
North West Territories	148 slides
Arctic Islands *	134 slides
Canada (Sets 1-11)	889 slides
East (Sets 1-4, 11)	427 slides
West (Sets 5-11)	596 slides

In addition to individual scenes, several colour mosaics of an entire region are available and a colour slide of the mosaic is included in the appropriate regional collection (sets with mosaics included are indicated by an asterisk in the list above).

The Continuing Education Administration of the Laboratory for Applications of Remote Sensing (LARS) [768] at Purdue University has published a minicourse series called Fundamentals of Remote Sensing since 1976. Some specialist knowledge is required for the course, which consists of about fifteen sets of approximately thirty slides each and an accompanying cassette tape that gives descriptions, asks questions and suggests exercises on the topics under consideration. The tape narrative is condensed in a companion booklet that contains imagery reproductions, tests, tables and graphic illustrations which should be used in conjunction with the audio-visual materials. Each minicourse may be used for individual tuition, or classwork if the booklet copies provided with each title are used. Although answers to the questions posed in each course are given on the respective tapes, a more in-depth treatment of the exercises is provided by the Instructor's Guides, which also contain slide hard-copies, maps and overlays. A guide to the complete range of minicourses which details the equipment necessary for proper use of the sets, together with course summaries and an indication of their respective levels of difficulty, is also available.

Pilot Rock Inc. [777] produce a large collection of 35mm slide-sets which make extensive use of vertical and oblique aerial photography as well as satellite imagery. Each set carries twenty to thirty slides and is accompanied by written notes compiled by expert researchers and academics. The level for which these sets are intended varies: some offer basic introductions, while others go further and contain more detailed and topic-specific information. An idea of the relevance of individual packs to different level groups is given in the company's Visual Teaching Aids booklet. A catalogue of sets that are available may also be requested from Pilot Rock Inc., who maintain an archive of nearly 40,000 images on a wide range of topics that include:

Skylab disciplinary set for agriculture
Mount St. Helens: A modern holocaust
U.S. Coastlines
Arid landforms
Land use patterns
Glaciology
Infrared high altitude photography
Introduction to digital processing of Landsat data

Several slide sets from this list are very educational, and the Mount St. Helens: A modern holocaust pack contains forty slides, posters and lecture notes. Both before and after shots are included at a variety of scales and in several different illustrative types, including aerial photography, MSS and weather satellite images. Other sets feature companion booklets, overlays, prints and user guides that present practical exercises.

The variety of visually appealing slide sets in the Earth from Space series, which are available from Focal Point Audio-Visual Ltd. [722], concentrate mainly on views of different regions and countries from space. There are twelve packs in this series and each set of twenty slides (forty for the British Isles) is accompanied by a small but very informative booklet that gives the date of imaging, image location and description of the slides' content. The packs, which come in hard-copy binders, are intended for geography teachers at school and college levels, and their coverage includes:

The British Isles: A view from space (double slide booklet)
Europe from space
The Middle East from space
Southern and Eastern Africa from space
Alaska and Canada from space
U.S. from space (various regional subdivisions)
Latin America and the Caribbean from space
Southern and S. E. Asia - China and Japan from space
The U.S.S.R.: A view from space
Australia: A view from space

Focal Point Audio-Visual Ltd. also publish several other slide sets, and two of these, Voyager in Space, Jupiter and its Moons and Voyager in Space, Saturn, present selected examples from the Voyager mission. One set, The Interpretation of Remotely Sensed Images, is issued with a forty-five minute tape and consists of forty slides that cover the physical basis of remote sensing and provides examples of images from both spaceborne and atmospheric sensors. Although ostensibly concerned with weather observation techniques, the twenty-slide pack on Observing and Forecasting the Weather contains some nice examples of imagery. Detailed explanatory notes and questions are also provided and the set appears to be aimed at grade-level students. A similar set of twelve slides, entitled Meteorology from Space, was jointly published in 1981 by Space Frontiers Ltd. [737] and the Royal Meteorological Society [736]. Space Frontiers Ltd. produce several other slide packs of relevance to remote sensing which have accompanying sound cassettes and illustrate a great deal of early space photography, together with some applications.

Each module of ten to twenty slides is accompanied by very detailed notes that fully describe the images. Several less relevant titles are offered and these deal with the early Viking, Voyager, Pioneer and Mariner missions, as well as the more up-to-date Skylab flights. A detailed overview of atmospheric circulation and weather patterns is provided by Weather Study with Satellites, a pack that contains over 100 slides and comes in three sets: Cyclones, Fronts and Anticyclones; Waves, Stratus, Fog and Sea Breezes; and Convection Patterns, Vortices and Global Views. The latter title is also available in filmstrip format with an accompanying cassette tape. A very detailed and comprehensive booklet of notes is provided and the text is supplied by R. R. Fotheringham, who also produced *The Earth's atmosphere viewed from space* [154], an excellent collection of annotated satellite images that provide clear examples of atmospheric systems.

The Technology Applications Center (TAC) [743] publishes a series of annotated slide sets which each contain twenty 35mm slides and a list of annotations. The Full and Partial Earth Set and the Lunar Set contain superb examples of Apollo mission photography. Orbital views of Mars, together with surface shots of the planet from the Viking landers, are included in the Mars/Viking Set. An artist's 'concept' pack, the Space Shuttle Set, illustrates uses of this launch vehicle and the related Launch Vehicles Set provides a retrospective look at developments in the U.S.A.'s space programme. Skylab products are listed exhaustively and there are seventeen sets of imagery that illustrate various regions and potential applications. Several remote sensing slide/tape shows and filmstrips are available at the time of writing but are probably being discontinued. The show titles currently offered include:

Food watch by satellite
Hydrology by satellite
Aerial and orbital remote sensing systems
Wildlife management and remote sensing
An introduction to remote sensing
Space settlement one

The latter two shows are also available on filmstrip and all of these sets are excellent for introducing remote sensing and its applications to a wide and non-specialist audience. An average of fifty-five colour 35mm slides, a narration on cassette tape and a script or study guide are contained in each show.

The International Institute for Aerospace Survey and Earth Sciences (ITC) [500] produces several very practical and innovative illustrative products that are suitable for training and higher education. Perhaps the most advanced set is the Stereo Slide Collection (unfortunately recently discontinued), which consists of over eighty superslides and detailed accompanying notes. Each slide has been set up so that it may be viewed stereoscopically; a modified projector is necessary, but the ITC can advise interested parties on where to obtain one. The slides are concerned with the interpretation of different-scale aerial-photographic images in the usual application divisions, and guidance, together with advice on the slides' use, is provided by the companion notes. Several slide and tape shows are available on subjects which range from the use of interpretative instruments to air-photo interpretation for different application areas. The application sets have between fifty and one hundred slides each and are accompanied by long tapes and sometimes workbooks, while the instrument packs usually contain thirty-five slides and the companion cassette comments on each of these for about one minute. A related, but much longer, series of several hundred slides that describe photogrammetric instruments, aerial cameras, film types, filter and non-photographic sensors is also available. It is a very useful visual catalogue of instrumentation, but does not feature many examples of remotely sensed imagery.

Several hundred slides grouped into sets of thirty to one hundred images that deal with various sensors, imagery types and applications are available from Remote Sensing Enterprises [785]. A Guide and Explanation for remote sensing slide sets that briefly describes each slide and provides an indication of how the images may be used in various exercises and applications is also published by Remote Sensing Enterprises. As this is available separately, interested teachers can decide how they may make best use of the slide sets prior to purchase. Each slide pack has expert acccompanying notes by F. Sabins, which are taken from undergraduate and postgraduate lecture courses presented at the author's university. Apart from being an excellent education resource, the packs may also be used in conjunction with the various publications available from Remote Sensing Enterprises. Of particular interest is a set of 100 slides which accompany the *Remote sensing laboratory manual* [461] and *Instructor's key for remote sensing laboratory manual* [460] and reproduce and illustrate the very comprehensive maps, diagrams, overlays, exercises and tasks outlined in the printed companions. The *Remote sensing laboratory manual* is in several sections, which include amongst others aerial photography, spacecraft imagery, Landsat imagery, radar, and digital image processing. Each slide is fully annotated, thus aiding the user in answering the various questions posed in the manual, although model answers to all questions and map exercises are present in the *Instructor's key for remote sensing laboratory manual*.

A brief, but colourful, introductory set of eighteen slides with an accompanying and informative pamphlet has been produced by the Education and Training Working Group of the National Remote Sensing Centre (NRSC) [703]. The set is useful as it presents some particularly excellent views, such as a GOES image of hurricane David, together with a wide range of imagery from platforms which include Landsat, HCMM, Nimbus 5 (ESMR) and 7 (CZCS), SIR-A, Skylab and Seasat. A similar set of twenty slides produced by ESA-EPO [498] in 1985 does not feature such a range of sensors, but includes up-to-date images from Landsat TM 4 and 5. The set is attractively packaged and has a one-page explanation that describes TM sensor characteristics and lists the geographic coverage of each slide, which range from Tromsö to the Bosporus and the Sea of Marmara.

An educational SPOT satellite image package is being offered by Nigel Press Associates Ltd. [729] of Edenbridge, Kent. Consisting of multispectral, panchromatic and stereoscopic prints, with scales between 1:50,000 and 1:100,000, together with a small slide set and colour anaglyph, the teaching aid provides a useful introduction to SPOT Image data.

A lot of the remote sensing consultancy and specialist firms listed in Part III have archives of slides on a very wide range of topics. They are usually pleased to assemble a customized slide set for individuals at quite reasonable prices, and can also offer annotations or commentaries.

It has already been mentioned that keeping up to date with audio-visual products is very difficult, although at the time of writing all of the resources listed here are still available. For any discontinued items it may be worth consulting the RESORS slide database described earlier in this section, which can loan or copy commercially available sets.

7.2 Videocassettes and Films

The EROS Data Center [742] holds a number of 16mm motion picture films that are available for two to three day loan within the United States subject to the borrower paying return postage. An overview of the role of the EROS Office and EROS Data Center is provided by the fourteen-minute EROS Response to a Changing World, produced in 1975. The film is oriented towards a general audience and briefly refers to applications of remote sensing technology. A similar movie, The National Domain (1979, 28 minutes), illustrates the work of the USGS, including the equipment they use and how the information obtained is applied in environmental studies and natural resource studies. A more technical explanation of how remote sensing is used in urban, crop, forest, wetland, grassland, geological, mineral and fuel exploration, flood, volcanic eruptions and disaster area applications is given in the film NASA: Earth Resources Technological Satellite (1973, 30 minutes). The Whole Earth's Invisible Colours (1973, 20 minutes) uses an interesting combination of film and music tones to explain the electromagnetic spectrum. Applications and potential applications of remote sensing are referred to and the film is an excellent primer for student groups.

The American Society of Photogrammetry and Remote Sensing (ASPRS) [740] has produced two films entitled Mineral Exploration: The Use of Remotely Sensed Data, and Vegetation Assessment: The Use of Remotely Sensed Data, in co-operation with the EROS Data Center. Both movies were produced in 1981, are twenty-five minutes long, and present a range of environmental issues in a way that makes them suitable for a general audience.

A large number of films produced by NASA are available on free loan from regional centres. A catalogue that offers a brief commentary on the films and lists film titles, their length and production date may also be obtained from the various distributors. In the U.K., the British Interplanetary Society [711] is NASA's regional representative and will furnish a list of the films it has for hire, or purchase, on request. A wide variety of technical and specialist topics are covered by the films and several from the NASA Environmental Series, 'Landsat: a Satellite for all Seasons' are particularly interesting. Remote Possibilities (1977, 14.5 minutes) provides a series overview and demonstrates the useful regional picture presented by Landsat. Three films, The Wet Look (1976, 14.5 minutes), The Fractured Look (1976, 14.5 minutes) and Pollution Solution (1976, 14.5 minutes), that cover a variety of applications were produced in 1976. The films, which deal respectively with water resources, mineral exploration and satellite monitoring of pollution damage, provide a dated, yet useful, introduction to the various applications of satellite imagery. Land for People, Land for Bears (1977, 14.5 minutes) and Growing Concerns (1977, 14.5 minutes) are in the same series and show how Landsat may be used in strategic land planning and agricultural resource monitoring and crop prediction. At a more advanced level is A Certain Distance (1970, 29 minutes); this film focusses on remote sensing applications in range management and forestry and was produced by the Pacific Southwest Forest and Range Experiment Station as a training aid for students and professionals.

A videotaped course entitled Advanced Remote Sensing Techniques is available for purchase or loan from the University of Arizona - Microcampus [793]. The course consists of fifteen videotapes, of approximately one hour each, that provide a very comprehensive outline of digital image processing and remote sensing techniques. The series authors are P. Slater, R. Hunt and R. Schowengerdt, who are respectively responsible for the sections on: Fundamentals of Digital

Processing of Remote Sensing Data (Tapes 1-11); Digital Image Processing (Tapes 12-13); and Digital Image Classification (Tapes 14-15).

Five videocassettes are available from the Continuing Education Administration [753] at Purdue University. The series provides both introductory and specialist material on the quantitative aspects of remote sensing and includes:

An introduction to quantitative remote sensing
Pattern recognition in remote sensing
Correction and enhancement of digital imagery
Numerical analysis in forecast management
The multispectral properties of soils

The course was produced in 1981 and each cassette lasts for about thirty minutes and has several sets of accompanying notes written by P. M. Swain and S. M. Davis. The series content reflects much of the material in the authors' book, *Remote sensing: the quantitative approach* [221], and is suited for postgraduates and higher-level students. Two videos that describe remote sensing applications were produced by the NRSC [703] in 1985. The twelve-minute Remote Sensing - Geological Applications is a broad introduction to remote sensing and is really only suitable for schools and colleges. Remote Sensing - Coastal and Marine Applications is aimed at the same audience and during its twenty-minute running time it outlines the potential contribution of remote sensing to coastal studies and at the same time provides a brief overview of available satellite systems. Of more value to researchers and higher-level students are Remote Sensing - Mineral Exploration (1980, 43 minutes) and Synthetic Aperture Radar (1981, 60 minutes), which present useful information on advanced techniques and applications and do not waste time with lengthy introductions to the basic concepts in remote sensing.

The ITC [500] have a small audio-visual unit that is mainly used to support the Institute's educational activities; however, a substantial archive of over 180 films and videocassettes, sixty slide sets and more than thirty audiocassettes and accompanying textbooks is also available. It is possible that some of the films may be borrowed if a special request is made to the ITC, but copies of some films are unavailable and the more relevant titles are probably in heavy demand.

Many other titles that cover a wide range of education and training skills are listed in the ITC CALT/AVC Resource Centre Catalogue. Several films that illustrate the operation of instruments and provide guides to learning skills have been prepared specifically by the audio-visual centre to aid in the specialized and very comprehensive courses that are offered at the ITC.

On the whole remote sensing videos are, with a few exceptions, quite basic introductions to the discipline, suitable only for for school or college consumption. Although films and videos publicize remote sensing and increase awareness of its uses, they suffer from being out of date and limited in choice. In most cases more topical information can be gleaned from the well-illustrated published literature that is available. New slide sets, films and videos are being issued all the time and apart from regular correspondence with those organizations that regularly produce audio-visual materials, the best way to keep informed of new titles is to consult the available online services.

The major machine-readable catalogue for audio-visual products is AV ONLINE, produced by Access Innovations Inc., and containing records back to 1913. This database may be searched through DIALOG and holds information on the producers, distributors, lengths and languages of over 375,000 16mm films, videos, filmstrips, slides and slide sets, overhead transparencies and audio tapes of educational value. A similar database, MARC FILMS, is offered by the Library of Congress (LOC) [770], and includes all film material of educational or institutional value released within the United States and Canada since 1972.

7.3 Remotely Sensed Image Maps

Of the satellite maps available, very few are useful for any detailed study because of scale considerations; they do however present an attractive and easy-to-relate-to view of the capabilities and synoptic coverage provided by satellite imagery. A select few go further and provide classifications and analyses of remote areas that are not well mapped in any other way. Satellite image maps are a little-publicized and poorly circulated resource, and although there are several hundred maps, photomaps and mosaics published, they appear to be poorly catalogued and often very few are held, even at important archival centres. Many of the maps available are based on Landsat or Skylab imagery, which may be geometrically corrected to cartographic coordinates and subsequently amalgamated with particular topographic features from ordinary maps. The map scales vary: some photomaps feature scales down to 1:50,000, while smaller scales of 1:250,000, 1:500,000 or even 1:1,000,000 are more common for satellite maps and mosaics.

A useful series of nearly 100 photomaps that cover Botswana, Chad, the Gambia and other areas in Africa are listed in *Africa: Guide to available mapping* [462], available from McCarta Ltd. [727]. McCarta produce several other catalogues of map publications including *Americas: Guide to available mapping* [463], *Europe: Guide to available mapping* [465] and *Asia, Australia, Pacific and Indian Oceans: Guide to available mapping* [464]. As well as topographic maps these very useful catalogues list over forty-five satellite and photomaps, together with information on their respective scale, publication date and coverage. The guide to the Americas is particularly good and identifies a number of image maps, including several published by the USGS. A variety of serial and non-serial satellite maps of the United States are available from the USGS Western Distribution Branch [798] and Eastern Distribution Branch [796]. The maps were produced from the early 1970's to the early 1980's from a selection of images that include black-and-white RBV or MSS colour and colour-enhanced mosaics. Several other experimental satellite image maps, or more accurately posters, of TM imagery over areas including Great Salt Lake and Washington DC have been issued by the USGS and are available either from it or the ASPRS [740].

Various U.S. Federal agencies, commercial organizations and societies publish satellite image maps. Perhaps the most outstanding of these is the U.S.

Department of Agriculture's (DOA) Soil Conservation Service [749], which produced an image map of the entire United States from Landsat-1 imagery. Over seventy-five accompanying image sheets, together with notes and overlays, were also published by the DOA in 1974 and are available at different scales for selected areas of the United States.

A series of forty map sheets [469-478] that cover a variety of developing and more remote countries including Burma, Bangladesh, Bhutan, Nepal, Peru and India is distributed by International Mapping Unlimited [766] for the World Bank [803]. These are probably the most useful set of satellite image maps currently available, as their geographical coverage relates to areas that are generally difficult to map accurately using conventional cartographic techniques. The interest level is high and unlike many of the more popular 'poster' images, these maps detail land cover and various land-type classifications. Several sheets cover most countries, and in addition to a useful set of fourteen maps that index Landsat availability for the areas given above under the title, *Landsat Index Atlas of the Developing Countries of the World* [493], there are nineteen newly published and annotated sheets of 1984 MSS imagery that cover parts of Nepal at a scale of 1:250,000. The Ordnance Survey's Overseas Directorate also produces a great deal of map material on developing countries and former British colonial territories. Most of the several hundred maps it issues each year are based on aerial photographs or photogrammetric measurements from these.

Many other countries publish satellite image maps on an irregular basis, but the availability of these is difficult to establish and the remote sensing centre, national mapping agency, or geological survey of countries of interest should be able to inform interested parties as to what satellite image maps they hold. Libraries of remote sensing centres that are concerned with education and training will usually have a collection of photomaps for teaching purposes. A good example of this practice is the Map Library of the Cartography Department at the ITC [500], which holds over forty maps that cover a wide variety of countries. Larger organizations such as the LOC Geography and Map Division [770] also hold substantial archives of cartographic material and have extensive acquisition lists and catalogues that detail their holdings.

An excellent source of data on map and aerial photographs is the *High school geography project: sources of information and materials, maps and photographs* [446], published in 1978 by the Association of American Geographers. Although somewhat out of date, this publication identifies some very useful and still current sources of information, despite being strongly biased towards U.S. aerial and space photography data sources. The guide is extremely comprehensive and includes sections that provide the contact addresses for photo-map and satellite-map suppliers.

An excellent and very comprehensive general guide to map sources is Hodgkiss and Tatham's *Keyguide to information sources in cartography* [466], published by Mansell in 1986. Most of the major organizations and publications that produce or list maps are included. Two of the most useful general information sources for cartographic material and related products are *Modern maps and atlases: an outline guide to twentieth century production* [468] and the *International Yearbook of Cartography* [467]. Further useful sources of up-to-date information on new map issues are the news and developments pages of the more cartographically inclined journals such as the *American Cartographer* [29], *Canadian Surveyor* [37], *Geodeziia i Kartografiya* [55] and *Surveying and Mapping* [119]. Many of the research projects mentioned in these, and other journals that are more specific to remote sensing, have had photomaps, image mosaics and related products prepared for them. These are potentially very valuable and often contain detailed small-scale maps, together with interpretations and a lengthy accompanying narrative. Locating and subsequently obtaining this type of small map is problematic as they are seldom indexed in the literature, however, several bibliographies list geographical areas of research and a study of these titles, combined with a review of the papers cited, may prove rewarding.

7.4 Atlases and Manuals of Remotely Sensed Imagery

Very few of the imagery atlases currently available are more than glossy picture books that are suitable only for light browsing or introductory-level familiarization with satellite images. Most of the examples mentioned here have well presented reproductions of images which are accompanied by some fairly basic explanatory text and an introductory resumé of how the images were formed, the principles behind this process and the current state-of-the-art in sensor technology at the time of publication. Unfortunately the images included in these atlases do not truly reflect the current state-of-the-art, as the production processes associated with publications of this kind are lengthy, and different sensors have usually been developed and are operational by the time the atlases are printed. For instance, at the time of writing, only a few atlases contain TM imagery, which has been available for some years, and none features SPOT images.

The earliest atlas to contain images from the first Landsat satellite is *Mission to Earth: Landsat views the world* [492], compiled with the help of an editorial team headed by Nicholas M. Short and published in 1976 by NASA. Several hundred images are presented and each has a brief descriptive narrative. Several atlases deal with early Apollo, Gemini and Skylab photography and two of these are notable. The first, *Earth Photographs from Gemini III, IV and V* [488], features images and accompanying maps on lining papers, whilst the second, *Skylab Explores the Earth* [489], is a more ambitious effort than the usual atlas, as it contains references to geophysics and remote sensing, addresses, brief essays and informative lecture-style entries.

A series of very useful atlases, or more accurately, image reference manuals, that focus on NASA's experimental satellite missions have been produced at the JPL. The SIR-A and B initiatives are covered by *Space Shuttle Columbia views the world with imaging radar - the SIR-A experiment* [484] and *Shuttle imaging radar views the Earth from Challenger - the SIR-B experiment* [485], whilst the Seasat mission is illustrated by *Seasat views North America, the Caribbean, and Western Europe with imaging radar* [483].

Two very popular atlases are *Earthwatch: a survey of the world from space* [490] and *Man on Earth: the works of man, a survey from space* [491]; both are 160 pages long and were published by Sidgwick & Jackson in 1981 and 1983 respectively. *Earthwatch* covers a number of geomorphological themes and presents excellent-quality images and accompanying descriptions in a basic introductory way, while *Man on Earth* follows the same format, but concentrates on man's influence on his environment.

Perhaps the best recent atlas is *Images of the world: an atlas of satellite imagery and maps* [487], which was published in 1984 by Collins as the English version of the German *Diercke Weltraumbild Atlas*. This book contains more maps than images. Some are in colour, which enables a useful comparison to be made between them and the 'true colour' Landsat images that are featured. The atlas is slightly let down by the fact that there are no TM images, but this detail is partly compensated for by the accompanying narrative, which suggests how the book may be used and provides an introduction to remote sensing.

Several atlas-style publications attempt to give an overview of a whole country. These include *Weltraumbild Atlas, Deutschland, Österreich, Schweiz* [480], but one of the better publications, not least for its coverage of little-known territories, is the *Atlas of false colour Landsat images of China* [479]. This three-volume bilingual (English and Chinese) set has 553 sheets that display Landsat 1, 2 and 3 images at a scale of 1:500,000. China's territory and border regions, as well as all of the islands in the South China Sea, are included as false-colour composites that display a variety of geomorphological and topographical features.

Images of Earth [486], published in 1984 by George Philip, provides examples of current satellite imagery, together with a retrospective series of attractive colour photographs from the Space Shuttle, Apollo and Gemini missions. The most recently published selection of Landsat and other sensor images is in *Britain from space: an atlas of Landsat images* [481]. This glossy book includes examples from TIROS-N AVHRR, SEASAT Synthetic Aperture Radar (SAR) and SPOT simulation, as well as Landsat, amongst the thirty-two images used to attain coverage of the whole British Isles. Unfortunately many of the images are from Landsat MSS, and tend to look rather poor owing to the large areas covered.

PART II

Annotated Bibliography of Sources of Information

Annotated Bibliography of Sources of Information

The aim of this annotated bibliography is to identify the key literature, services and products that are available for remote sensing, together with the means of keeping up to date with these services. The listing is by no means an exhaustive one, but hopefully the selected items are some of the most fundamental in remote sensing research. The author would be very pleased to receive regular updates on all remote sensing materials from organizations, publishers and individuals.

Certain abbreviations have been used throughout this section, including A (annual), M (monthly), bi-M (bi-monthly), Q (quarterly), I (irregular), p.a. (per annum), and variations thereof.

Research in Remote Sensing

1 **Ansari, M. B.** and **Newman, L. P.** *Nevada directory of maps and aerial-photo resources.* Santa Cruz, California: Western Association of Map Libraries. Occasional paper No. 11, 1984. 164pp. ISBN 0-939112-13-2

2 **Archbold, T., Laidlaw, J. C.** and **McKechnie, J.,** consultant eds. *Engineering Research Centres.* Harlow: Longman, 1984. 1,031pp. ISBN 0-582-90018-2

A primary reference source providing outlines and profiles of over 5,000 commercial, industrial and academic research centres. Arrangement is alphabetical and by country and there are establishment title and subject indexes. Over sixty entries on remote sensing and aerial photography are included.

3 *British Reports, Translations and Theses.* British Library Document Supply Division, Publications Section, Boston Spa, Wetherby, West Yorkshire LS23 7BQ, England

As well as announcing newly received theses from British universities, this periodical mirrors the reports collection held at the British Library Document Supply Division (which includes numerous NASA reports), and covers translations that may be of interest to those involved in remote sensing under the headings of Aeronautics, Earth Sciences, Electronic and Electrical Engineering, Computer Science and Space Technology.

4 *Catalogue of European Industrial Capabilities in Remote Sensing*
A. A. Balkema, P.O. Box 1675, NL-3000 BR Rotterdam, The Netherlands
A useful directory of European activities in remote sensing that is divided into four parts. Part 1 provides descriptions of hardware and equipment offered by European companies, while Part 2 displays a table of the services available from these companies. Part 3 summarizes individual companies' research focus in remote sensing, and Part 4 comprises up-date sheets in the form of a mini-questionnaire. The up-date facility is particularly interesting as the information returned from it is entered onto the Commission of the European Communities' videotex service, thereby providing an excellent current awareness service.

5 *Comprehensive Dissertation Index 1861-1972*
University Microfilms International, 300 North Zeeb Road, Ann Arbor, MI 48106, U.S.A.
This index lists all of the dissertations received at American universities during the above period, as well as some of those received at Canadian and foreign academic institutions.

6 **Cracknell, A.** and **Hayes, L.** *Remote Sensing Yearbook* A
London: Taylor & Francis, 1987. 700pp. ISBN 0-85066-378-4
An excellent publication that contains a comprehensive directory section on remote sensing organizations, many of which have not been previously documented. The introductory narrative is however disappointing, as it mainly consists of reprints of papers that have been published elsewhere. The most recent edition of this book is most helpful for current awareness as it contains a bibliography of the previous year's key journal publications in remote sensing.

7 *Current Research in Britain: Physical Sciences* A
Boston Spa: The British Library Document Supply Centre. ISBN 0-7123-2027-X
This publication replaces *Research in British Universities, Polytechnics and Colleges (RBUPC)*, produced by the British Library since 1979. A very useful means of keeping up to date with UK academic research in remote sensing and related sciences.

8 *Directory of UK Space Capabilities*
London: HMSO, 1987. 250pp. ISBN 0-1151-3981-1
A directory of the space-related activities of more than 200 manufacturers, agencies and universities that lists government departments, companies, consultants, trade associations, universities, polytechnics, institutes and learned societies. Quick-reference tables ensure the rapid location of goods and services.

9 *Dissertation Abstracts International, Section B: Physical Sciences and Engineering.* University Microfilms International, 300 North Zeeb Road, Ann Arbor, MI 48106, U.S.A.

This formidably large abstracting series provides a guide to the vast collection of doctoral dissertations that are available. Each dissertation entered in this publication has been microfilmed, and is available from UMI in this format.

10 Dowman, I. J., ed. *Directory of research and development activities in the United Kingdom in land survey and related fields.* 3rd edn. London: Surveyors Publications, 1982.

A very useful list of organizations that are either involved in research, or which occasionally carry out research. Other groups are also identified, and these include equipment manufacturers, their agents, learned societies, professional bodies, and some commercial institutions.

11 *EARSeL Directory.* updated-A

EARSeL Secretariat, 148 rue du Fg. Saint Denis, P.O. Box 60, F-75462 Paris Cedex 10, France

A superb and very comprehensive directory of the activities, facilities, services and products of the EARSeL membership. Nearly 180 member laboratories are listed in full directory style, whilst over fifty addresses of observer laboratories are included. The form of publication is very innovative; each year a member will receive up-date sheets that report on developments with particular organizations, and these replace the previous entries for those laboratories, which are easily removed because of the directory's ring-bound construction. A summary is included at the front, and this details each country in turn, and provides information on the numbers of members, observers, temporary observers and approximate number of researchers in every country. An index of members, observers and temporary observers gives the name of each group, and a code which refers to where that entry may be found.

12 *European Research Centres.* 6th edn. bi-A

Harlow: Longman, 1986. 2,453pp.

A directory of organizations in science, technology, agriculture and medicine. Arranged alphabetically by country with subject and establishment title indexes. Updated versions are expected to appear every two years, and there are over one hundred entries on aerial photography and remote sensing.

13 Fitch, J. M., consultant ed. *Earth and Astronomical Sciences Research Centres.* Harlow: Longman, 1984. 762pp. ISBN 0-582-90020-4

Easy access is provided to over 3,500 research centres and details of personnel and programmes are given. With over 150 entries on remote sensing and aerial photographic organizations and their activities, alphabetically listed by country, this is the most useful, and specific to remote sensing, of the numerous Longman guides.

14 *French remote sensing activities, services and products.* A
Prospace, 2 place Maurice Quentin, F-75039 Paris Cedex 1, France
English and French versions of this catalogue are available, and it is published in several volumes. The catalogue provides up-to-date news on French space activities and is particularly concerned with the SPOT and Ariane programmes. The individual may keep informed of the catalogue's information, between successive editions, by referring to the Prospace association's tri-annual publication, *News from Prospace.*

15 Harvey, A. P., ed. *European Sources of Scientific and Technical Information* bi-A
Harlow: Longman, 1985. 655pp. ISBN 0-582-90029-8
Over 1,500 entries divided into general and discipline-specific chapters. Establishment title and subject indexes.

16 *Index to Theses Accepted for Higher Degrees by the Universities of Great Britain and Ireland, and the Council for National Academic Awards* semi-A
Aslib, Publications Department, Information House, 26-27 Boswell Street, London WC1N 3JZ, England
Subject and author indexes.

17 *Industrial research in the United Kingdom.* 11th edn. A
Harlow: Longman, 1985. 655pp. ISBN 0-582-90029-8
Published every two years, with over 3,000 research profiles, including remote sensing and related fields. This directory has a personnel name index of over 7,000 senior staff in industry, as well as establishment title and subject indexes.

18 *Jane's Spaceflight Directory 1986* A
London: Jane's Publishing, 1986. 453pp. ISBN 0-2106-0367-3
Information on international space programmes; excellent accounts of those in Europe and the U.S.A., particularly on Landsat and the Space Shuttle. Very comprehensive satellite launch tables (pp. 422-446) for 1982-1985. An excellent section gives details on world space centres, and there is a listing of space contractors.

19 Jaques Cattell Press, ed. *Industrial Research Laboratories of the United States.* 19th edn. A
New York: R. R. Bowker, 1985. 742pp. ISBN 0-8352-2070-2
Nearly 11,000 entries that cover a wide range of organizations. Alphabetical listing of research laboratories, together with State, personnel and classification indexes. Also available online.

20 The Remote Sensing Association of Australia. *National directory of remote sensing in Australia.* 2nd edn. New South Wales: Remote Sensing Association of Australia, 1980. 206pp.

A directory of those groups and individuals involved in remote sensing in Australia, with division by subject. Probably quite out of date. Keyword and organization indexes are included, together with a list of RSAA membership.

21 *Research in geography: a catalogue of doctoral dissertations 1976-1986*
University Microfilms International, 300 North Zeeb Road, Ann Arbor, MI 48106, U.S.A.

22 *Research in meteorology and climatology: a catalogue of doctoral dissertations 1975-1986*
University Microfilms International, 300 North Zeeb Road, Ann Arbor, MI 48106, U.S.A.

23 *The United Kingdom Remote Sensing Directory 1987*
National Remote Sensing Centre (NRSC), R 190 Building, Space Department, Royal Aircraft Establishment, Farnborough, Hampshire GU14 6TD, England
This long-awaited up-date of the Department of Trade and Industry's *Remote Sensing of Earth Resources* is very comprehensive in its coverage of academic, commercial, industrial, consultancy, governmental and learned society involvement in remote sensing. The directory is indexed into the sections described above, and there are over 200 entries, which describe the facilities, services and courses offered by various organizations, together with lists of staff involved in these aspects of remote sensing and an indication of their geographical area or areas of interest. There are some particularly useful entries which tabulate the types and availability of remotely sensed data for different regions of the UK.

24 **United Nations.** *Education, Training, Research and Fellowship Opportunities in Space Science and Technology and Its Applications: A Directory*
United Nations Department of Political and Security Council Affairs, 1986. pp 67-161.
Although the section on remote sensing and related disciplines is confined to the pages indicated above, this directory is a useful guide to education and training in a variety of UN countries. Over thirty countries are represented, many by more than one education and training facility.

Journals

25 *ACSM Bulletin* bi-M
American Congress on Surveying and Mapping, 210 Little Falls Street, Falls Church, VA 22046, U.S.A.

26 *Acta Astronautica* M
Pergamon Press Inc., Journals Division, Maxwell House, Fairview Park, Elmsford, NY 10523, U.S.A.

27 *Advances in Space Research* I
Pergamon Press Inc., Journals Division, Maxwell House, Fairview Park, Elmsford, NY 10523, U.S.A.
for
Committee on Space Research, 51 boulevard Montmorency, F-75016 Paris, France
A very high-quality hardback publication detailing the latest developments in space and remote sensing research via a series of collected papers. Occasional special editions are devoted entirely to one aspect of remote sensing.

28 *Aerial Archaeology* Q
Aerial Archaeology Publications, 15 Colin McLean Road, East Dereham, Norfolk NR19 2RY, England
Aerial reconnaissance, photography and photo-interpretation for archaeology are the main subjects presented in this journal's papers.

29 *American Cartographer* bi-A
American Congress on Surveying and Mapping, 210 Little Falls Street, Falls Church, VA 22046, U.S.A.
The technical articles in this journal are dominated by cartographical content; however, remote sensing is strongly represented in many cases. Sections include notes, literature surveys, an advertiser index and software and book reviews.

30 *American Society of Cartographers Bulletin* Q
American Society of Cartographers, Box 1493, Louisville, KY 40201, U.S.A.

31 *Applied Optics* semi-M
American Institute of Physics, 335 E. 45th Street, New York, NY 10017, U.S.A.
for
Optical Society of America (OSA), 1816 Jefferson Place, N.W. Washington, DC 20036, U.S.A.
Largely theoretical in content; many of the techniques and ideas discussed are either directly or indirectly of relevance to remote sensing. An interesting section that provides patent abstracts, together with some drawings, is included at the front of the journal. A rapid communications feature is available for the publication of new research results and there are also sections on OSA news and membership.

32 *Applied Optics. Supplement* I
American Institute of Physics, 335 E. 45th Street, New York, NY 10017, U.S.A.

33 *Australian Journal of Geodesy, Photogrammetry and Surveying* semi-A
University of New South Wales, School of Surveying, P.O. Box 1, Kensington, N.S.W. 2033, Australia

Technical papers on remote sensing from the south-west Pacific region are given publication preference. Notices and publication lists are detailed.

34 *Bildmessung und Luftbildwesen* bi-M
Herbert Wichmann Verlag, Rheinstr. 122, Postfach 210949, D-7500 Karlsruhe 21, Federal Republic of Germany
One of the oldest and best-established European journals. Articles, in German with English and French abstracts, are focussed on the photogrammetric and aerial photographic nature of remote sensing.

35 *Bollettino di Geodesia e Scienze Affini* Q
Istituto Geografico Militare, Via C. Battisti 10, I-50100 Florence, Italy

36 *Canadian Journal of Remote Sensing* 2 p.a.
Canadian Aeronautics and Space Institute, 60-75 Sparks Street, Ottawa, Ontario K1P 5A5, Canada
for
Canadian Remote Sensing Society, 222 Somerset Street West, Suite No. 601, Ottawa, Ontario K2P 0J1, Canada
A wide range of learned articles are presented in English with French abstracts.

37 *Canadian Surveyor / Géomètre Canadien* Q
Canadian Institute of Surveying, Box 5378, Sta. F, Ottawa, Ontario K2C 3J1, Canada
A high-quality publication with book reviews and news articles that is primarily of interest to those researchers concerned with the surveying and mapping side of remote sensing.

38 *Cartographica* Q
University of Toronto Press, Toronto, Ontario M5S 1A6, Canada

39 *Centro Interamericano de Fotointerpretación. Revista* A
Centro Interamericano de Fotointerpretación, Carrera 30 No 47a-57, Apdo. Aereo 53754, Bogotá, Colombia

40 *Commercial Space* Q
McGraw-Hill Inc., 1221 Avenue of the Americas, New York, NY 10020, U.S.A.
A great many of the articles and papers presented in this journal are specifically for persons with professional and commercial interests in space science and technology. Businesslike features on remote sensing are included, many of which contain complex and comprehensive information on policies, finances and economics.

41 *Computer Vision, Graphics and Image Processing* M
Academic Press Inc., Journals Division, 111 Fifth Avenue, New York, NY 10003, U.S.A.
This journal features original research papers on the processing of pictorial information, reviews, application-oriented articles, short notes, and replies to and comments on papers published in previous issues.

42 *Danmarks Tekniske Hoejskole Instituttet for Landmaaling og Fotogrammetri, Meddelelse* I
Danmarks Tekniske Hoejskole, Instituttet for Landmaaling og Fotogrammetri, Landmaaling 7, DK-2800 Lyngby, Denmark

43 *DFVLR Jahrebücher* A
Deutsche Gesellschaft für Luft- und Raumfahrt e.V., Godesberger Allee 70, D-5300 Bonn 2, Federal Republic of Germany

44 *Earth Oriented Applications of Space Technology* Q
Pergamon Press Inc., Journals Division, Maxwell House, Fairview Park, Elmsford, NY 10523, U.S.A.

45 *Earth Resources Mapping in Africa* Q
Regional Centre for Services in Surveying, Mapping and Remote Sensing, Regional Remote Sensing Facility, Box 18118, Nairobi, Kenya

46 *East African Geographical Review* A
Uganda Geographical Association, Makerere University, Box 7062, Kampala, Uganda
Limited aerial photographic content.

47 *Environmental Monitoring and Assessment* Q
D. Reidel Publishing Company, Box 17, NL-3300 AA Dordrecht, The Netherlands
Devoted to progress in pollution control. Special editions occasionally focus on remote sensing techniques in pollution monitoring.

48 *ESA Journal* Q
European Space Agency, 8-10 rue Mario Nikis, F-75738 Paris Cedex 15, France
This large-format publication is available free and contains both specialized and general articles on a range of space technology subjects. There are comprehensive listings of ESA-sponsored symposia and ESA publications. A regular list of patents applied for by ESA is included, as are brief abstracts of external publications by ESA staff. Mainly in English, with occasional contributions in French.

49 *Fotogrametría, Fotointerpretación y Geodesia* I
Sociedad Mexicana de Fotogrametría, Fotointerpretación y Geodesia, Apartado Postal 25-447, Mexico 13, D.F., Mexico
Of interest to professionals in photogrammetry, remote sensing and geodesy.

50 *Fotogrammetriska Meddelanden/Photogrammetric Notes* I
Royal Institute of Technology, Department of Photogrammetry, S-100 44 Stockholm, Sweden
An irregular series of brief technical papers on photogrammetry in Swedish, English and German.

51 *Fotointerpretacja w Geografii* I
Universytet Slaski w Katowicach, Ul. Bomkowaa 14, 40-007 Katowice, Poland
In Polish with English and French summaries.

52 *Geocarto* Q
Geocarto International Centre, GPO Box 4122, Hong Kong
First published in 1986, this excellent journal has a high print quality and carries informed contributions on a wide range of remote sensing techniques, these being complemented by useful news and review sections.

53 *Geodézia és Kartográfia* bi-M
Cartographia, P.O. Box 132, 1443 Budapest, Hungary
Although strongly biased toward geodesy, this Hungarian-language publication features papers, abstracted in English and German, with some remote sensing and geographic information system related information. Society news and reviews of pertinent professional literature are included.

54 *Geodeziia i Aerofotos'emka* bi-M
Ministerstvo Vysshego i Srendnego Spetsial'nogo Obrazovaniia S.S.S.R; Izdatel'stvo Instituta Inzhenerov Geodezii, Aerofotos'emki i Kartografii, Moscow 8, U.S.S.R.
A Russian-language journal with an emphasis on geodesy, but still containing some useful photogrammetric articles.

55 *Geodeziia i Kartografiya* M
Glavnoe Upravlenie Geodezii i Kartografii pri Sovete Ministrov, Moscow, U.S.S.R.
A dominantly geodetic Russian-language periodical that features occasional articles on the cartographic applications of remote sensing.

56 *Geodezja i Kartografia* Q
Polska Akademia Nauk, Komitet Geodezji, Panstwowe Wydawnictwo Naukowe, Ul. Miodowa 10, 00-251 Warsaw, Poland

A limited remote sensing content is apparent in this journal, which is largely concerned with geography and cartography.

57 *Geografia* Q
Instituto de Geociências e Ciências, UNSEP, Rua 10, Caixa Postal 178, Rio Claro, Brazil

58 *Geografisch Tijdschrift* 5 p.a.
Koninklijk Nederlands Aardrijkskundig Genootschap, Weteringschans 12, NL-1017 SG Amsterdam, The Netherlands
Technical articles, news, publication reviews and advertisements.

59 *Geo-Processing* Q
Elsevier Scientific Publishing Company, Box 211, NL-1000 AE Amsterdam, The Netherlands
Contains learned papers and surveys that detail research and development in the field of spatial information systems. Data collection, structure, analysis and display are covered.

60 *IEEE Transactions on Aerospace and Electronic Systems* bi-M
IEEE Aerospace and Electronic Systems Society, Institute of Electrical and Electronics Engineers Inc., 345 E. 47th Street, New York, NY 10017, U.S.A.

61 *IEEE Transactions on Geoscience and Remote Sensing* bi-M
IEEE Geoscience and Remote Sensing Society, Institute of Electrical and Electronics Engineers Inc., 345 E. 47th Street, New York, NY 10017, U.S.A.
This publication is perhaps one of the most technical of the remote sensing journals; most of the papers are largely theoretical and many have a very strong physics and electronics bias. As well as the technical content, this journal features book reviews, news, and a calendar of forthcoming events. Recommended reading for the scientist who wishes to be acquainted with the forefront of remote sensing research and development.

62 *IEEE Transactions on Microwave Theory and Techniques* M
IEEE Microwave Theory and Techniques Society, Institute of Electrical and Electronics Engineers Inc., 345 E. 47th Street, New York, NY 10017, U.S.A.

63 *IEEE Transactions on Pattern Analysis and Machine Intelligence* bi-M
Institute of Electrical and Electronics Engineers Inc., 345 E. 47th Street, New York, NY 10017, U.S.A.

64 *IEE Proceedings Part F: Communications, Radar and Signal Processing* bi-M
Institution of Electrical Engineers, Savoy Place, London WC2R 0BL, England

65 *IEE Proceedings Part H: Microwaves, Antennas and Propagation* bi-M
Institution of Electrical Engineers, Savoy Place, London WC2R 0BL, England

66 *IMPACT of Science on Society* Q
United Nations Educational, Scientific and Cultural Organization, 7 place de Fontenoy, F-75700 Paris, France
Has special issues on remote sensing (e.g. No. 140 Vol. 35. No. 4 1985: Examining a blue-green gem) with excellent picture reproduction. A good selection of papers from a markedly international audience are presented. A list of UNESCO publication distributors is featured. Published in French, Chinese, Russian and Korean.

67 *International Journal of Geographical Information Systems* Q
Taylor & Francis Limited, Rankine Road, Basingstoke, Hampshire RG24 0PR, England
First published in 1987. The appearance of this journal illustrates the growing interest in this specialist field and provides a forum for technical discussion on geographic information systems, many of which involve remote sensing to some degree.

68 *International Journal of Remote Sensing* bi-M
Taylor & Francis Limited, Rankine Road, Basingstoke, Hampshire RG24 0PR, England
for
Remote Sensing Society, Department of Geography, University of Nottingham, University Park, Nottingham NG7 2RD, England
The official journal of the Remote Sensing Society; a variety of technical and review papers and letters are featured, together with comprehensive news sections, a calendar of forthcoming events, and a summary of remote sensing projects in progress.

69 *Israel Department of Surveys. Photogrammetric Papers* I
Department of Surveys, P.O. Box 14171, Tel Aviv 611141, Israel
In Hebrew with English abstracts.

70 *Issledovanie Zemli iz Kosmosa* bi-M
Akademiia Nauk S.S.S.R., Izdatel'stvo Nauka, Moscow, U.S.S.R.
The major Russian-language remote sensing journal, which features a table of contents in English. A cover-to-cover translation is available as the *Soviet Journal of Remote Sensing*.

71 *ITC Journal* Q
International Institute for Aerospace Survey and Earth Sciences, P.O. Box 6, NL-7500 AA Enschede, The Netherlands

A very well produced journal with well balanced contributions on remote sensing as well as photogrammetry and aerial surveys. Features include book reviews, conference reports, news, notices, an international calendar, and thesis and journal abstracts. Special issues are devoted to a specific topic in remote sensing, generally application oriented.

72 *Itogi Nauki i Tekhniki: Seria Geodeziia i Aeros'emka* I
Vsesoyuznyi Institut Nauchno-Tekhnicheskoi Informatsii (VINITI), Baltiiskaya ul. 14, Moscow A-219, U.S.S.R.

73 *Itogi Nauki i Tekhniki: Geodeziya i Kartografiya* I
Vsesoyuznyi Institut Nauchno-Tekhnicheskoi Informatsii (VINITI), Baltiiskaya ul. 14, Moscow A-219, U.S.S.R.

74 *Jena Review* Q
VEB Carl Zeiss Jena, Carl-Zeiss-Strasse 1, 6900 Jena, German Democratic Republic
A high-quality and large-format journal that contains some useful technical information on photogrammetry, at the same time promoting and advertising developments in research and hardware.

75 *Journal of Geophysical Research* bi-M
American Geophysical Union, 2000 Florida Avenue, N.W., Washington DC 20009, U.S.A.

76 *Journal of Imaging Technology* bi-M
Society of Photographic Scientists and Engineers, 7003 Kilworth Lane, Springfield, VA 22151, U.S.A.
This serial features technical, tutorial and review papers that occasionally mention remote sensing, usually aerial photography. Short technical notes are communicated, together with abstracts from other journals in the same sort of field.

77 *Journal for Photogrammetrists and Surveyors* I
Jenoptik Jena GmbH, Carl-Zeiss-Strasse 2, 69 Jena, German Democratic Republic
Highlights instrument and research developments at Carl Zeiss Jena. Published in English, French and German.

78 *Journal of the Atmospheric Sciences* I
American Meteorological Society, 45 Beacon Street, Boston, MA 02108, U.S.A.

79 *Journal of the Environmental Satellite Amateur Users Group* Q
2512 Arch Street, Tampa, FL 33607, U.S.A.

Provides a means by which amateurs can obtain technical information on environmental satellite data.

80 *Journal of the Indian Society of Photointerpretation and Remote Sensing (Photonirvachak)* Q
Indian Institute of Remote Sensing, National Remote Sensing Agency, No. 4 Kalidas Road, P.O. Box 135, Dehra Dun 248 001, Uttar Pradesh, India
Papers are in English and generally from contributors within India. They tend to focus on practical applications of remote sensing in agriculture and land-based studies.

81 *Journal of the Japan Society of Photogrammetry and Remote Sensing* Q
Japan Society of Photogrammetry and Remote Sensing (JSPRS), Daiichi Honan Building, 601 2-8-17, Minami-ikebukuro, Toshima-ku, Tokyo, Japan
Scholarly papers in Japanese with occasional English abstracts, reviews of symposia and general news.

82 *Journal of Wuhan Technical University of Surveying and Mapping* I
Department of Photogrammetry and Remote Sensing, Wuhan Technical University of Surveying and Mapping, 23 Lo-Yu Road, Wuhan, Hubei Province, People's Republic of China

83 *Landinspektöeren* Q
Teknisk Forlag A-S, Skelbaekgade 4, DK-1717 Copenhagen V, Denmark

84 *Maanmittaus* semi-A
Maanmittaustieteiden Seura r.y., Pl 85, SF-00521 Helsinki 52, Finland
Finnish papers with English and German abstracts, primarily of interest to photogrammetrists and surveyors.

85 *Mapping Sciences and Remote Sensing* Q
V. H. Winston and Sons Inc., 7961 Eastern Avenue, Silver Spring, MD 20910, U.S.A.
also (co-publishers) American Congress of Surveying and Mapping, American Society of Photogrammetry
This journal contains direct translations of papers selected from about twelve established Soviet and Eastern European periodicals on geography, photogrammetry, aerial photography and remote sensing. Also published are state-of-the-art reviews, polemics, special features and translations of manuscripts and papers that are forthcoming from other sources. Expensive, owing to translation costs.

86 *Memoires de Photo-Interpretation* I
Ecole Pratique des Hautes Etudes, Librairie Touzout, 38 rue Saint Sulpice, F-75278 Paris, Cedex 06, France

87 *Nachrichten aus dem Karten- und Vermessungswesen* I
Institut für Angewandte Geodäsie, Abteilung Photogrammetrische Forschung, Richard-Strauss-Allee 11, D-6000 Frankfurt am Main 70, Federal Republic of Germany
Carries articles on cartographic surveying and photogrammetry.

88 *Nigerian Engineer* Q
Nigerian Society of Engineers, Technical Committee, 360 Herbert Macaulay Street, P.O. Box 1041, Yaba, Nigeria

89 *Nigerian Geographical Journal* semi-A
Ibadan University Press, University of Ibadan, Ibadan, Nigeria
Only a small proportion of the papers appearing in this journal are of interest to remote sensors; however, it is one of the only African publications including any remote sensing at all.

90 *Nigerian Journal of Photogrammetry and Remote Sensing* I
Nigerian Society of Photogrammetry and Remote Sensing, ECA Regional Centre for Training in Aerial Surveys, Department of Photointerpretation and Remote Sensing, Department of Photogrammetry, PMB 5545, Ile-Ife, Oyo State, Nigeria
Although very irregular in publication, this is, to the best of the author's knowledge, the only journal that deals directly with remote sensing available within Africa.

91 *NOAA* Q
U.S. National Oceanic and Atmospheric Administration, Office of Public Affairs, Rockville, MD 20852, U.S.A.

92 *Ocean-Air Interactions: Techniques, Observations and Analyses, an International Journal* Q
Gordon & Breach, Marketing Department, P2 INTERN, P.O. Box 786, Cooper Station, New York, NY 10276, U.S.A.
A new journal, first published in 1986, that aims to report on the latest techniques in ocean dynamics and global climate research. Articles and short communications are published, together with special issues on satellite instrumentation and software.

93 *OEEPE* I
Institut für Angewandte Geodäsie, Abteilung Photogrammetrische Forschung, Richard-Strauss-Allee 11, D-6000 Frankfurt am Main 70, Federal Republic of Germany
for
Organization Européenne d'Etudes Photogrammétriques Expérimentales, 350 boulevard 1945, P.O. Box 6, NL-7500 AA Enschede, The Netherlands

Although dominantly concerned with photogrammetry, this journal also features papers and news on remote sensing and aerial photography.

94 *Optical Engineering* bi-M
Society for Photo-Optical Instrumentation Engineers, P.O. Box 10, Bellingham, WA 98227, U.S.A.
Each edition of this journal focusses on a specific topic, and around two-thirds of the articles are on this, whilst the remainder are on other subjects. Details of future special issues are provided, as well as book reviews, Society news, an advertiser index and a calendar of forthcoming Society events and meetings.

95 *OZ Österreichische Zeitschrift für Vermessungswesen und Photogrammetrie* Q
Österreichischer Verein für Vermessungswesen und Photogrammetrie, Friedrich-Schmidt-Platz 3, A-1082 Vienna, Austria
Contributions vary from technical papers to learned historical reviews, both abstracted in English. Book reviews and advance announcements of conferences are also included.

96 *Pattern Recognition* bi-M
Pergamon Press Inc., Journals Division, Maxwell House, Fairview Park, Elmsford, NY 10523, U.S.A.
for
Pattern Recognition Society, National Biomedical Research Foundation, Georgetown University Medical Centre, 3900 N.W. Reservoir Road, Washington, DC 20007, U.S.A.
Scholarly articles on pattern recognition and digital image processing are a feature of this journal.

97 *Pattern Recognition Letters* bi-M
North-Holland Publishing Company, Box 211, NL-1000 AE Amsterdam, The Netherlands
for
International Association of Pattern Recognition (IAPR), Department of Physics Astronomy, University College London, Gower Street, London WC1E 6BT, England
Contributions are short, enabling the journal to publish a timely and concise range of articles on the developments and literature of pattern recognition.

98 *Photogrammetria* bi-M
Elsevier Scientific Publishing Company, Box 211, NL-1000 AE Amsterdam, The Netherlands
for
International Society for Photogrammetry and Remote Sensing (ISPRS), Department of Photogrammetry, S-100 44 Stockholm, KTH, Sweden

This publication, the official journal of the ISPRS, contains a wide range of technical articles on different applications of remote sensing and photogrammetry. There are also sections concerned with book reviews, reports and news.

99 *Photogrammetric Engineering and Remote Sensing* M

American Society for Photogrammetry and Remote Sensing, 210 Little Falls Street, Falls Church, VA 22046, U.S.A.

With a monthly circulation of over 10,000 copies this is perhaps one of the best-known and most popular remote sensing journals. Contributions are of the highest quality and come from a diverse and international range of authors. Society news and political developments in remote sensing are well represented and the journal carries a great deal of advertising material. As from 1987 the journal's format will be increased in size to A4, in order to accommodate more illustrative material and more usefully display remote sensing information.

100 *Photogrammetric Journal of Finland* 1-2 p.a.

Finnish Society of Photogrammetry, Institute of Photogrammetry, Helsinki University of Technology, SF-02150 Espoo 15, Finland

This journal deals almost exclusively with photogrammetry. The technical papers and features are in English, German and Finnish.

101 *Photogrammetric Record* semi-A

Photogrammetric Society, Department of Photogrammetry and Surveying, University College London, Gower Street, London WC1E 6BT, England

The official journal of the Photogrammetric Society. Contains learned articles that have a strong bias toward photogrammetry; however, the photogrammetric side of remote sensing is well represented. As well as containing book reviews, notes and membership lists, this journal offers an excellent abstracting service entitled 'Photogrammetry around the World', which presents brief abstracts from papers on photogrammetry and remote sensing from a wide range of periodical and conference literature.

102 *Photointerprétation* bi-M

Editions Technip, 27 rue Ginoux, F-75737 Paris Cedex 15, France

The principal aim of this journal is to act as a teaching aid for the interpretation of remotely sensed images. The content focusses on case studies or methodological syntheses, possibly dealing with a sensor, method, geographic region or specific topic. To facilitate interpretation and the use of the superb colour plates, overlays and stereo pairs, presentation is in the form of a folder with removable loose-lea sheets. Papers are brief and in French, with an extended summary in English and another language, often Italian or German.

103 *Radio Science* bi-M

American Geophysical Union, 2000 Florida Avenue, N.W. Washington, DC 20009, U.S.A.

This journal includes a section on electromagnetic phenomena related to physical problems, and this occasionally features highly technical papers on the application of electromagnetic technology to remote sensing of the environment.

104 *Remote Sensing of Environment* bi-M
Elsevier Science Publishing Company Inc. (New York). 52 Vanderbilt Avenue, New York, NY 10017, U.S.A.

105 *RESTEC Journal* I
Remote Sensing Technology Centre of Japan, Uni-Roppongi Building, 7-15-17, Roppongi, Minato-Ku, Tokyo 106, Japan

106 *Revista Cartográfica* semi-A
Instituto Panamericano de Geografía e Historía, Servicios Bibliograficos, Ex-Arzobispado 29, Col. Observatorio, Deleg. Miguel Hidalgo, 11860 Mexico, DF., Mexico
Although this semi-annual review publication is cartographically biased, at least half of the articles are on photogrammetry and remote sensing. Contributions are in English, French, Spanish and Portuguese.

107 *Revista Teledetección* A
Centro de Levantamientos Intergrados de Recursors Naturales por Sensores Remotos - Clirsen, Subgerencia de Teledetección, Calle Paz y Mino s/n Edf. I.G.M., 4 to. piso/Casilla 8216, Quito, Ecuador

108 *Rivista del Catasto e dei Servizi Technici Erariari* semi-A
Ministero delle Finanze, Largo Leopardi No 5, I-00185 Rome, Italy
for
Istituto Poligrafico dello Stato, Piazza Verdi 10, I-00198 Rome, Italy

109 *Società Italiana di Fotogrammetria, Topografia e Teledetection. Bollettino* Q
Società Italiana di Fotogrammetria, Topografia e Teledetection, Piazzale R. Morandi 2, I-20121 Milan, Italy
The official journal of the Società Italiana di Fotogrammetria, Topografia e Teledetection, which features scholarly articles and news in Italian, with occasional English and French contributions.

110 *Société Belge de Photogrammétrie, Télédétection et de Cartographie. Bulletin Trimestriel* Q
H. Van Olffen, C.A.E. Tour Finances, Boîte No. 38, 50 boulevard du Jardin Botanique, B-1010 Brussels, Belgium
The official journal of the Société Belge de Photogrammétrie, Télédétection et Cartographie. In Flemish with English, German and French summaries.

111 *Société Française de Photogrammétrie et de Télédétection. Bulletin* Q
Société Française de Photogrammétrie et de Télédétection, 2 avenue Pasteur, F-94160 Saint Mandé, France
The official journal of the Société Française de Photogrammétrie et de Télédétection. Papers are in French with English and German abstracts. Contains book reviews and society notes, including brief biographies of leading figures in photogrammetry and remote sensing.

112 *South African Journal of Photogrammetry, Remote Sensing and Cartography* semi-A
South African Society for Photogrammetry, Remote Sensing and Cartography, P.O. Box 69, Newlands 7725, South Africa
News, book reviews and a well balanced selection of papers on remote sensing and photogrammetry.

113 *Soviet Journal of Remote Sensing* bi-A
Harwood Academic Publishers, 50 W. 23rd Street, New York, NY 10010, U.S.A.
A cover-to-cover translation of the Russian journal *Issledovanie Zemli iz Kosmosa.* Technical papers and abstracts of forthcoming articles. Expensive, owing to translation costs.

114 *Space* bi-M
The Shephard Press Limited, 111 High Street, Burnham, Buckinghamshire SL1 7JZ, England
A controlled-circulation publication that is free to established professionals in the aerospace industry. This periodical contains features and articles on political, commercial and research developments in space technology and remote sensing, including some excellent reviews of major space programmes and initiatives.

115 *Spaceflight* M
British Interplanetary Society, 27/29 South Lambeth Road, London SW8 1SZ, England
Quite a lot of remote-sensing-related material is included in this journal, which also covers Society news, book reviews and letters.

116 *Space Horizons* Q
Pakistan Space and Upper Atmosphere Research Commission (SUPARCO), Remote Sensing Applications Centre (RESACENT), P.O. Box 3125, Karachi-29, Pakistan

117 *Space Markets* Q
Interavia S.A., 86 avenue Louis-Casai, P.O. Box 162, CH-1216 Cointrin, Geneva, Switzerland

With a more relaxed editorial style than the 'hard core' research journals, this publication encourages the ready assimilation of its interesting articles on space research development and commercialization.

118 *Space Policy* Q

Butterworth Scientific Limited, P.O. Box 163, Guildford, Surrey GU2 5BH, England

An excellent publication that is divided into several sections. Opinions are expressed about specific aspects of space policy and informed articles discuss issues in space, many involving remote sensing. The book reviews are complemented by news, including extracts from official documents and statements.

119 *Surveying and Mapping* Q

American Congress on Surveying and Mapping, 210 Little Falls Street, Falls Church, VA 22046, U.S.A.

Primarily of interest to surveyors and cartographers. Remote sensing is included as it has applications in both these fields.

120 *Surveying Engineering* I

American Society of Civil Engineers, 345 East 47th Street, New York, NY 10017, U.S.A.

Remote sensing applications in engineering are a feature of this journal.

121 *Survey Ireland* A

Survey Department, College of Technology, Bolton Street, Dublin 1, Republic of Ireland

122 *Surveying News* I

VEB Carl Zeiss Jena, Department for Surveying and Photogrammetrical Instruments, Carl-Zeiss Strasse 1, 69 Jena, German Democratic Republic

This journal features technical papers concerned with photogrammetry and surveying.

123 *Survey Review* Q

C. F. Hodgson & Son Limited, Unit 4, Central Trading Estate, Staines, Middlesex TW18 4UR, England

Largely devoted to surveying although some very technical articles on photogrammetry are included.

124 *Vermessung, Photogrammetrie, Kulturtechnik/Mensuration, Photogrammétrie, Génie Rural* M

Offset Haus AG, Postfach, CH-8021 Zurich, Switzerland

The papers in this journal are in either French or German and feature abstracts in the other language on photogrammetry and photogrammetric surveying. There is a technical news section and a title listing of papers appearing in other journals.

125 *Vermessungstechnik, Zeitschrift für Geodäsie, Photogrammetrie und Kartographie der DDR* M
VEB Verlag für Bauwesen, Französische Strasse 13-14, 1086 Berlin, German Democratic Republic
Learned technical, historical and review papers biased towards photogrammetry and cartography are included, some of which contain useful information on remote sensing. Papers are abstracted at the front and back in German, Russian and English. A guide to other Eastern-bloc publications on remote sensing, photogrammetry and cartography is presented, as well as news, conference reports and reports on photogrammetric instruments.

126 *Vision* I
Context Vision AB, Linkoping, Sweden
A new publication intended for the commercial market.

127 *Zeiss Information* I
Carl Zeiss, Postfach 1369/1380, D-7082 Oberkochen, Federal Republic of Germany
This trade-related serial publicizes developments in instrumentation at Carl Zeiss and presents useful articles on the photogrammetric applications possible with their equipment.

128 *Zeitschrift für Flugwissenschaften und Weltraumforschung* bi-M
Deutsche Forschungs- und Versuchsanstalt für Luft- und Raumfahrt e.V., Linder Hoche, Postfach 906058, D-5000 Cologne 90, Federal Republic of Germany
Published in German and English, this journal features original papers and review articles on all aspects of flight and space science, including aerospace research and technical fields.

129 *Zeitschrift für Vermessungswesen* M
Verlag Konrad Witter KG, Nordbahnhofstrasse 16, Postfach 147, D-7000 Stuttgart 1, Federal Republic of Germany
Technical articles and reviews of surveying and photogrammetry. Symposia and book reports.

Textbooks and Monographs

Aerial Photography

130 *Airborne reconnaissance VI (SPIE Vol. 354).* 1982. 169pp
The International Society for Optical Engineering, P.O. Box 10, Bellingham, WA 98227, U.S.A.
A comprehensive report on the results of involved discussions on the role and status of aircraft remote sensing, sensor utilization and the technology and concepts of airborne reconnaissance.

131 *Airborne reconnaissance VII (SPIE Vol. 424)*. 1984. 197pp.
The International Society for Optical Engineering, P.O. Box 10, Bellingham, WA 98227, U.S.A.
This meeting's volume covered reconnaissance equipment, image interpretation, technology and annotation data handling and recording, in four thorough sessions.

132 *Airborne reconnaissance VIII (SPIE Vol. 496)*. 1984. 187pp.
The International Society for Optical Engineering, P.O. Box 10, Bellingham, WA 98227, U.S.A.
The most recent volume in the *Airborne reconnaissance* series concentrates on sensors and ancillary equipment, technological advances, development and testing, and intelligence extraction and exploitation.

133 Avery, T. E. *Forester's guide to aerial photographic interpretation.* Washington D.C.: U.S.D.A. Forest Service, 1969. 40pp.
Although dated, this handbook is very useful for the forester with no remote sensing experience and provides a concise and easy-to-understand review of the major photo-interpretation techniques.

134 Avery, T. E. and **Graydon, L. B.** *Interpretation of aerial photographs.* 4th edn. Minneapolis: Burgess, 1985. 554pp. ISBN 0-8087-0096-0
A comprehensive guide to aerial photographic interpretation and the technology and techniques behind it, including a useful section on the basic principles of digital image processing. A guide to available photography is provided, together with a report on the Aerial Photography Summary Record System and the National High Altitude Photography Programme (NHAP). Colour and black-and-white plates are extensively featured and there is a glossary, as well as references at the end of each chapter.

135 Burnside, C. D. *Mapping from aerial photographs.* London: Collins, 1985. 348pp. ISBN 0-00-383036-5
This mathematical text, which is largely concerned with the photogrammetric side of aerial photography, provides an extremely comprehensive view of this science. Appendices have photogrammetric and technical information, and there is a useful bibliography.

136 El-Ashry, M. T., ed. *Air photography and coastal problems.* Strondsberg, Pennsylvania: Dowdes, Hutchinson & Ross, 1977. 425pp. ISBN 0-87933-252-2
This volume is a collection of articles that date back to 1934, and cover the use of air photographs for monitoring shoreline changes, coastal geology and in solving coastal engineering problems. A somewhat dated list of air photography sources is provided, together with extensive bibliographies and author citation and general indexes.

137 Howard, J. A. *Aerial photo-ecology.* London: Faber & Faber, 1970. 325pp. ISBN 0-571-08592-X

Much of this work is still pertinent since it concerns basic interpretation techniques and although a general overview, it gives major emphasis to forest areas. Informative bibliographies are included.

138 Lo, C. P. *Geographical applications of aerial photography.* Newton Abbot: David & Charles, 1976. 330pp. ISBN 0-844-80872-5

This basic text, for geographers with limited photogrammetric experience, demonstrates the potential and limitations of aerial photographs. Examples, which are drawn from physical and human geography, are supported by references at the end of each chapter.

139 Nagao, M. and **Matsayama, T.** *A structural analysis of complex aerial photographs.* New York and London: Plenum Press, 1980. 199pp. ISBN 0-306-40571-7

This well presented book provides a unique view of the structure of aerial photographs, especially in suburban areas. The technical approach to aerial photographic interpretation is explained and the information given is augmented by excellent plates, figures and tables. A useful introduction on photographic interpretation with the aid of artificial intelligence, expert systems and digital techniques is given.

140 Paine, D. P. *Aerial photography and image interpretation for resource management.* New York: John Wiley & Sons, 1981. 571pp. ISBN 0-471-01857-0

An excellent elementary textbook on aerial photo-interpretation that includes some photogrammetry, together with an introduction to opto-mechanical scanning systems. Each chapter begins with a specific set of objectives and concludes with sample questions and problems based on the objectives. Colour plates usefully illustrate the examples, and there are references after each section.

141 Smith, J. T., ed. *Manual of color aerial photography.* Falls Church, Virginia: American Society of Photogrammetry, 1968. 550pp.

A complete technical overview of this important field of remote sensing. Comprises numerous contributions from experts in the various disciplines involved, from the physics of colour, through to the applied aspects of colour photography. Presentation is excellent as well as being concise and systematic. Although dated in its approach to enhancement techniques, as well as film/filter combinations presently available, this is still a highly recommended text for those wishing to understand more about the background to this precise subject area. Sets of useful tables and charts, a glossary, list of abstract and a list of colour system tables/chips are also featured.

142 Wear, J. F., Poper, R. B. and **Orr, P. W.** *Aerial photographic techniques for estimating damage by insects in western forests.* Washington D.C.: U.S.D.A. Forest Service, 1966. 79pp.

This text provides a basic introduction to this tried and tested field, although not helped by a lack of photographic examples. Bibliographies are included.

Agriculture

143 American Society of Photogrammetry. *Symposium on remote sensing for vegetation damage assessment.* Falls Church, Virginia: American Society of Photogrammetry, 1978. 548pp.

The primary source for this particular study area, which provides thirty-two papers introducing the reader to damage assessment, as well as several seminal review papers from established experts in this field.

144 Berg, A., ed. *Application of remote sensing to agricultural production forecasting.* Rotterdam: A. A. Balkema, 1981. 266pp. ISBN 90-6191-089-7

A useful combination of information on crop production, crop-yield weather modelling and the contribution of remote sensing techniques to these fields. Some rather detailed concepts are dealt with in the eighteen papers contained in this monograph.

145 Blasquez, C. H. and **Horn, F. W.** *Aerial colour infrared photography: applications in citriculture.* Scientific and Technical Information Branch: NASA Reference Publication 1067, 1980. 82pp.

An excellent and well illustrated guide to this application, that ably illustrates the practicality of remote sensing. Intelligent usage of photographs and overlays demonstrates the proposed system's potential, and this text could be used as a model by which all future handbooks that wish to present practical applications might be judged. An appendix on remote sensing systems is included, as well as a glossary and references.

146 Deepak, A. and **Rao, K. R.** *Application of remote sensing for rice production.* Hampton: A. Deepak Publishing, 1984. 449pp. ISBN 0-937194-03-4

The proceedings of a Symposium that was held in Secunderabad, India (1981). The thirty-five invited papers that are featured cover the use of both satellite and aerial imagery for rice production monitoring and modelling.

147 Fraysse, G. *Remote sensing application in agriculture and hydrology.* Rotterdam: A. A. Balkema, 1980. 502pp. ISBN 90-6191-081-1

A comprehensive collection of thirty papers, both general and relating to specific examples, that are slightly biased towards agricultural applications of remote sensing technology.

148 Godby, E. A. and Otterman, J. *The contribution of space observations to global food information systems.* Oxford: Pergamon Press, 1978. 202pp. ISBN 0-08-022418-0

This monograph, Volume 2 in the COSPAR series Advances in Space Research, contains twenty-six papers that cover a little-acknowledged, but effective use of remotely sensed imagery, bringing together expertise on food production, distribution and space technology. Many of the problems and theories discussed are particularly applicable to the Third World.

149 White, L. P. *Aerial photography and remote sensing for soil survey.* Oxford: Clarendon Press, 1977. 104pp. ISBN 0-19-854509-6

Although dwelling on the remote sensing background, this book provides a good general introduction to its field.

Atmosphere

150 Barrett, E. C. and Martin, D. W. *The use of satellite data in rainfall monitoring.* London: Academic Press, 1981. 340pp. ISBN 0-12-079680-5.

A summary and preview of the remote sensing methods that are used to augment conventional rainfall data. Bibliographies.

151 Bolle, H. J. *Remote sounding of the atmosphere from space.* Oxford: Pergamon Press, 1979. 256pp. ISBN 0-08-023419-4

Again in the excellent series of COSPAR proceedings, this book provides an overview of this broad field of study and great emphasis is placed on retrieval methods and accuracy analysis.

152 Deepak, A. *Inversion methods in atmospheric remote sounding.* New York: Academic Press, 1977. 622pp. ISBN 0-12-208450-0

Twenty-one papers that cover the mathematical theory of inversion methods, as well as the application of these methods to various atmospheric phenomena. Discussions of the concepts and problems outlined are included, together with references at the end of each paper.

153 Deepak, A. *Remote sensing of atmospheres and oceans.* New York: Academic Press, 1980. 641pp. ISBN 0-12-208460-8

A collection of twenty-five papers discussing the state of knowledge and the results from research ranging from remote sounding of atmospheric temperature to ocean colour experiments. Divided into eight sections of about three papers each, which are: recent developments, atmospheric temperature sounding, interpretation of aerosol sounding, gases, remote sounding by microwaves, wind sounding, ocean parameter sounding and recent results from space. References are provided at the end of each paper.

154 Fotheringham, R. R. *The Earth's atmosphere viewed from space.* Dundee: Dundee University, 1979. 71pp.

An excellent collection of annotated satellite images providing some clear pictorial examples of atmospheric systems. A most useful and informative teaching guide that includes a list of recommended books.

155 Fymat, A. L. and **Zuev, V. E.** *Remote sensing of the atmosphere: inversion methods and applications.* Amsterdam: Elsevier Scientific Publishing Company, 1978. 327pp. ISBN 0-444-41748-6

A selection of twenty-five papers that give a broad overview of temperature, composition and particulate sounding of the atmosphere. References follow each paper and there are subject and author indexes.

156 Henderson-Sellers, A. *Satellite sensing of a cloudy atmosphere: observing the third planet.* London: Taylor & Francis, 1984. 340pp.

A highly readable overview of the problems encountered in atmospheric remote sensing.

157 Houghton, J. T., Taylor, F. W. and **Rodgers, C. D.** *Remote sounding of atmospheres.* Cambridge: Cambridge University Press, 1984. 343pp. ISBN 0-521-24281-9

This text provides a description of the main sensors and concepts involved in atmospheric remote sensing. There are chapters on remote sensing of our own atmosphere, as well as that of other planets in the solar system. Although the book is rather technical in parts, the main ideas are well illustrated and clearly explained.

158 Optical Society of America. *Topical meeting on optical techniques for remote probing of the atmosphere.* Optical Society of America, 1983.

A digest of over ninety papers on atmospheric investigations, many of which are directly or indirectly related to remote sensing. Extensive references are featured at the end of each chapter.

Civil Engineering and Planning

159 Kennie, T. J. M. and **Matthews, M. C.** *Remote sensing in civil engineering.* Glasgow and London: Surrey University Press, 1985. 357pp. ISBN 0-903384-48-5

Based on a short course, 'The Use of Remote Sensing in Civil Engineering', this book reviews the theoretical background to the subject, as well as providing case examples as illustrations of the available techniques. Contact addresses give details of how to obtain aerial photographs and satellite imagery and there is a glossary of remote sensing terms and a list of abbreviations.

160 Kiefer, R. W., ed. *Civil engineering applications of remote sensing.* New York: American Society of Civil Engineers, 1980. 193pp. ISBN 0-87262-253-3

The proceedings of a conference on civil engineering applications in remote sensing. The presented papers are specific examples of remote sensing applied as a solution to civil engineering problems. A guide to locating remote sensing information through the National Cartographic Information Centre is included. References are provided at the end of each paper and there are subject and author indexes.

Digital Image Processing

161 Bernstein, R. *Digital image processing in remote sensing.* New York: Institute of Electrical and Electronics Engineers Inc., 1978. 473pp. ISBN 0-87942-105-3

A selection of technical papers and articles that provide a foundation for digital image processing, covering all the stages of image manipulation from registration to extraction. A glossary, bibliography and a somewhat dated but still useful list of sources of earth observation data are all featured.

162 Ekstrom, M. P. *Digital image processing techniques.* London: Academic Press, 1984. 372pp. ISBN 0-12-236760-X

An advanced overview of the main image enhancement, restoration and compression techniques, that includes an interesting section on image processing systems.

163 Green, W. B. *Digital image processing a systems approach.* New York: Van Nostrand Reinhold, 1983. 192pp. ISBN 0-442-28801-8

A practical text with an emphasis on keeping theory to a minimum. Many of the examples are drawn from remote sensing. References are provided at chapter ends and there are appendices on data sources, imagery, software and documentation, together with a bibliography and list of further information sources.

164 Hord, R. M. *Digital image processing of remotely sensed data.* London: Academic Press, 1982. 256pp. ISBN 0-12-355620-1

This text is part of a comprehensive series on computer techniques, and this particular volume covers a wide range of image processing operations. Dated as regards the types of imagery included, but still providing a lucid account of the basic background procedures. A list of organizations, courses, image processing and interpretation services is provided, as are subroutine listings and a glossary.

165 Jensen, J. R. *Introductory image processing: a remote sensing perspective.* Englewood Cliffs, New Jersey: Prentice-Hall, 1986. 379pp. ISBN 0-13-500828-X

An excellent introduction that deals with modern image processing methodologies and processes in a practical manner. The basic concepts, considerations and techniques are outlined with a minimum of the usual detailed and often distracting technical information. References follow each chapter and there are appendices on digital image processing programmes and address lists of public and commercial suppliers of digital image processing hardware and software.

166 Niblack, W. *An introduction to digital image processing.* Denmark: Prentice-Hall International, 1986. 215pp. ISBN 0-13-480674-3

A concise introduction to this vast field of interest. The book is designed as a university-level teaching aid for those with some mathematical experience; hence the basic mathematical concepts behind each technique are summarized to good effect, and presented with practical examples in all fields of image processing. The topics covered include image filtering and segmentation, Fourier transform, geometric operations and classification, as well as image display. A useful discussion and problem section has been included at the end of each chapter, and overall the book fulfils its promise as an introductory text to the software scientist; however, applications scientists are recommended to search elsewhere.

167 Richards, J. A. *Remote sensing digital image analysis, an introduction.* Heidelberg: Springer-Verlag, 1986. 281pp. ISBN 0-387-16007-8

Based on a graduate course in digital image processing and analysis techniques, this publication covers basic and more involved algorithms. The educational parentage is illustrated by the provision of exercises at the end of each chapter which provide a useful summary without dictating the format of the text. As well as these problems, a useful bibliography is provided at each chapter ending, an intelligent addition to any publication of this kind. Regarding specific topics, the book starts with an introduction to data sources commonly encountered in digital form and progresses through geometric/radiometric enhancement techniques, image transformations and classification algorithms. Some useful appendices are also provided which outline some of the mathematics and statistics behind the observed methodologies. All sections are well planned (with the inclusion of 151 clear illustrations) and although somewhat involved in places the book fulfils the stated aim of providing an introduction to this field of remote sensing.

168 Schowengerdt, R. A. *Techniques for image processing and classification in remote sensing.* London: Academic Press, 1983. 249pp. ISBN 0-12-628980-8

An excellent overview of image processing is presented, together with a useful introduction to classification techniques. Remote sensing and image processing bibliographies are featured, together with details of digital image data formats, a review of look-up table algorithms, examination questions, and references at the end of each chapter.

Ecology

169 Fuller, R. M., ed. *Ecological mapping from ground, air and space.* Cambridge: Institute of Terrestrial Ecology, 1983. 142pp. ISBN 0-904282-71-6
Fourteen papers, with references, covering a wide range of ecological themes and demonstrating technical advances in the collection, processing and interpretation of ecological data.

170 Natural Wildlife Federation. *Application of remote sensing data to wildlife management.* Natural Wildlife Federation, Pecora IV: Scientific and Technical Series 3, 1978. 397pp. ISBN 0-912186-30-5
A collection of fifty-seven papers dealing with a wide range of habitat and species types. Biased towards the American experience, although the models are widely applicable for further surveys in related fields.

Forestry

171 Murtha, P. A. and **Harding, R. A.** *Renewable resources management: applications of remote sensing.* Falls Church, Virginia: American Society of Photogrammetry, 1984. 774pp. ISBN 0-937294-51-9
A collection of some seventy-eight papers presented at a symposium in 1983. Although ostensibly concerned with renewable resources, there is a bias towards vegetation monitoring and in particular forestry. A useful overview of vegetation damage assessment is presented.

Geography

172 Marble, D. F., Calkins, H. W. and **Peuquet, D. J.** *Basic readings in geographic information systems.* Williamsville, New York: SPAD Systems Limited. 325pp. ISBN 0-913913-00-6
Aimed at researchers and practitioners in any field who wish to know more about this wide-ranging topic. Excellently formatted, this publication follows an explanatory line from 'The Nature of a GIS' to 'GIS Design and Evaluation'. Bibliographies are included and there is an index at the end of each chapter.

173 Townshend, J. R. G. *Terrain analysis and remote sensing.* London: George Allen & Unwin, 1981. 231pp. ISBN 0-04-551036-9
A general introductory text that is mainly for geographers. Some excellent case studies concerning erosion monitoring and tropical soil survey are presented, and there are references after each chapter.

174 Verstappen, H. Th. *Remote sensing in geomorphology.* Amsterdam: Elsevier Scientific Publishing Company, 1977. 214pp. ISBN 0-444-41086-4
An excellent guide to geomorphological remote sensing; some sections require updating, but those on interpretation remain valid and are well illustrated with

photographs, stereo pairs and occasional colour plates. References are provided at the end of each chapter.

Geology

175 Carter, W. D., Rowan, L. C. and Huntington, J. F., eds. *Remote sensing and mineral exploration.* Oxford: Pergamon Press, 1980. 173pp. ISBN 0-08-024438-6

This proceedings volume presents a collection of twenty-five papers that reflect the international 'state of the art' in geological sciences and exploration for mineral and energy resources.

176 El-Baz, F. *Deserts and arid lands.* The Hague: Martinus Nijhoff, 1984. 222pp. ISBN 90-247-2850-9

A collection of eleven papers describing the utility of aerial photography and Landsat data in the study of semi-arid, arid and hyper-arid terrains. References follow each paper and provide a key to further reading in the specific areas that come under discussion.

177 Drury, S. A. *Image interpretation in geology.* London: Allen & Unwin, 1987. 243pp. ISBN 0-04-550037-1

An excellent and very up-to-date book that covers all aspects of geological remote sensing extremely well. Excellent colour plates and illustrations accompany the informative text, which features particularly interesting sections on digital image processing, thermal imagery and the recognition of rock types. Appendices concentrate on stereometry, image correction and sources of remotely sensed images, and there are general suggestions for further reading, together with more specific chapter-by-chapter bibliographies.

178 Fabbri, A. G. *Image processing of geological data.* Victoria: Van Nostrand Reinhold, 1984. 244pp. ISBN 0-442-22536-9

An excellent overview of image processing methodologies applied to geology that contains extensive appendices which detail dedicated digital image processing systems, geological image analysis programme packages, parallel-processing algorithms for minicomputers and a demonstration sequence of GIAPP. References are given at the end of each chapter and a glossary of terms unfamiliar to geologists is also provided.

179 Kronberg, P. *Photogeologie.* Stuttgart: Ferdinand Enke Verlag, 1984. 268pp. ISBN 3-432-94161-7

A German-language publication that provides a comprehensive guide to photogeology. Excellent-quality illustrations and an extensive and superb bibliography that provides a guide to monograph and serial literature on photogeology are included.

180 Miller, V. C. *Photogeology*. New York: McGraw-Hill, 1961. 248pp.
Although dated, this large-format text nevertheless presents a useful guide to photogeological methods. The mechanics and principles of photogeology are reviewed and complemented by a section of illustrations (which are annotated in the appendix) and exercises. Stereopairs are included of a variety of geological terrains. A good basic educational book with a comprehensive bibliography.

181 Siegal, B. S. and **Gillespie, A. R.** *Remote sensing in geology*. New York: John Wiley & Sons, 1980. 702pp. ISBN 0-471-79052-4
A comprehensive and practical text for geology students, which is divided into four sections, concerned with the physical basis of remote sensing, optical and digital data processing, interpretation, and remote sensing applications in geology. The stereo photographic content could be higher and the recent advances in remote sensing tend to make this text dated. A glossary is provided and there are references at the end of each chapter.

182 Von Bandat, H. F. *Aerogeology*. Houston: Gulf Publishing Company, 1962. 350pp.
An introduction to the practical use of aerial photography in geological investigations, with particular emphasis given to tectonic and lithologic interpretation of geomorphic units and landforms. A well structured text that is practically produced to aid stereo-interpretation of aerial photographs. References are cited at the bottom of each page.

Ice and Snow

183 Hall, D. K. and **Martinec, J.** *Remote sensing of ice and snow*. London: Chapman & Hall, 1985. 189pp. ISBN 0-412-25910-9
A very readable overview of this field that describes the interactions between oceanic, atmospheric, hydrospheric and cryospheric processes, and examines the various sensors and platforms used in remote sensing. References at the end of each chapter.

Introductions to Remote Sensing

184 Barrett, E. C. and **Curtis, L. F.** *Introduction to environmental remote sensing*. 2nd edn. New York: Chapman & Hall, 1982. 352pp. ISBN 0-412-23080-1
Divided into two sections; the first covers physical aspects of remote sensing, data acquisition, analysis, processing and interpretation, while the second details a wide range of applications. An excellent basic and introductory-level text that perhaps takes the reader further than the other books aimed at this area and backs up the information given with an intelligent bibliography.

185 Colwell, R. N., ed. *Manual of remote sensing.* 2nd edn. Falls Church, Virginia: American Society of Photogrammetry and Remote Sensing, 1983. 2,724pp. ISBN 0-937294-52-7

This two-volume set has been called the 'bible of remote sensing', and there are very few individuals who would disagree with this description. It is an essential reference work for anybody who is involved in remote sensing or related fields, and presents a very complete overview of all aspects of these sciences. The book is divided into thirty-six chapters, with Volume 1 covering the theoretical and system side of remote sensing, whilst Volume 2 is more concerned with the practical applications of a great variety of remotely sensed data. A review and preview of remote sensing systems is provided and deals with a range of systems, from optical imagers to radar and fluorosensors. The substantial narrative is accompanied by 220 full colour pages, over 250 colour figures, 2,072 illustrations, and extremely comprehensive bibliographies. Absolutely essential reading.

186 Curran, P. J. *Principles of remote sensing.* New York: Longman, 1985. 282pp. ISBN 0-582-30097-5

Now accepted as the main introductory text for undergraduate and graduate courses in remote sensing, this publication covers the whole range of remote sensing systems, as well as the background and physical nature of the science. Most of the past and present remote sensing systems are included, and there is also a lengthy section on the physical basis of remote sensing.

187 Holz, R. K. *The surveillant science: remote sensing of the environment.* 2nd edn. New York: John Wiley & Sons, 1985. 413pp. ISBN 0-471-08638-X

The update of an earlier overview that was published before Landsat had made its mark. With one exception all the articles are new to this edition, and each chapter covers a different imaging wavelength category and provides case studies to illustrate its respective use in a range of application areas.

188 Sabins, F. F. *Remote sensing: principles and interpretation.* 2nd edn. Oxford: W. H. Freeman & Co., 1986. 592pp. ISBN 0-7167-1793-X

A superb new edition of this work, which contains chapters on Fundamental considerations, Aerial photographs and multispectral images, Manned satellite images, Landsat images, Thermal infrared images, Radar images, Digital image processing and a range of application topics. Over 550 illustrations, including 16 colour plates, together with appendices, a glossary and references.

189 Szekielda, K. H. *Satellite remote sensing for resources development.* London: Graham & Trotman, 1986. 232pp. ISBN 0-86010-805-8

This book provides a comprehensive outline of the major remote sensing systems currently available, and the application areas in which they may be of use.

Land Use

190 Eden, M. J. and **Parry, J. T.**, eds. *Remote sensing and tropical land management*. London: John Wiley & Sons, 1986. 365pp. ISBN 0-471-90889-4

A useful and practical overview of this field, featuring a range of contributions on management, technology, classification and evaluation, survey and monitoring, thus presenting a balanced view of tropical land management. An interesting and contemporary section on the prospects for remote sensing in the tropics is provided and there are references following each section.

191 Johannsen, C. J. and **Sanders, J. L.**, eds. *Remote sensing for resource management*. Iowa City: Soil Conservation Society of America, 1982. 665pp. ISBN 0-935734-08-2

A well balanced overview of this wide field that contains fifty-five varied papers on topics ranging from strip mine reclamation to planning transmission line routes. A list of sources of remotely sensed data is given, as is information on image processing and interpretation services, image processing and hardware systems, and image processing and software systems. The Stateside Natural Resource Information System is explained and acronyms and a glossary are included.

192 Lindgren, D. T. *Land use planning and remote sensing*. Dordrecht: Martinus Nijhoff, 1985. 176pp. ISBN 90-247-3083-X

A useful text that introduces land planners to remote sensing, focussing mainly on aerial photographic interpretation and dealing largely with the analysis of small areas. Only established applications are presented and no preview is given of how remote sensing may develop into a useful tool for planning. The principles of remote sensing are explained in a way that is very useful for planners and those with an interest in the urban environment.

193 Ryerson, R. A. *Land use information from remotely sensed data*. Ottawa: Canada Centre for Remote Sensing, Users' Manual 80-1, 1980.

Methods of obtaining various types of land use data are explained, and a table of land use information available from remote sensing data is presented. Names of firms and contact addresses are included, as well as an extensive reference list.

194 Schanda, E. *Remote sensing for environmental sciences*. Heidelberg: Springer-Verlag, 1976. 367pp.

An introduction to remote sensing for natural scientists that emphasizes the newer methods, such as laser, radar and sonar remote sensing. Chapter length seems dependent on how developed the new technology is, perhaps at the expense of demonstrating useful applications. References are featured at the end of each chapter and several technical appendices are included.

Microwave

195 Allan, T. D. *Satellite microwave remote sensing.* Chichester: Ellis Horwood, 1983. 526pp. ISBN 0-85312-494-9
A comprehensive guide to the relative specifications and application capabilities of the sensors carried on Seasat, which will appeal to the experienced oceanographer.

196 Trevett, J. W. *Imaging radar for resources surveys.* London: Chapman & Hall, 1986. 313pp. ISBN 0-412-25520-0
An applications-oriented introduction to microwave technology, this publication pulls together the various radar systems now available and presents their characteristics and uses in a readable form. The first section of the book, which is involved with the acquisition and processing of these data, is in a form that is easy for the layman to understand and follow. A further section is devoted to applications of radar imagery for geological interpretation, forestry and land use. The advantage of this publication is its cross-discipline considerations and the ability to incorporate examples into a description of a particular system (airborne or satellite). The final section, dealing with satellite systems, covers both present and future platforms such as ERS-1 and Radarsat and relates the radar story up to the present moment. At the end of the book an invaluable bibliography is provided covering all aspects of the subject and these augment the reference sections at the end of each chapter.

197 Ulaby, F. T., Moore, R. K. and **Fung, A. F.** *Microwave remote sensing: active and passive, volume 1, MRS fundamentals and radiometry.* Reading, Massachusetts: Addison-Wesley, 1981. 456pp. ISBN 0-201-10759-0
This guide is suitable both as a general reference text and for researchers, scientists and graduate programmers. Areas covered include theory and techniques, modelling, instruments and applications. References are provided at the end of each chapter and three technical appendices and one on abbreviations, acronyms, system and satellite names are provided.

198 Ulaby, F. T., Moore, R. K. and **Fung, A. F.** *Radar remote sensing and surface scattering and emission theory.* Reading, Massachusetts: Addison-Wesley, 1982. 620pp. ISBN 0-201-10759-0
This volume, the second in this series on microwave remote sensing, details the fundamentals of surface scattering and emission for land and ocean surfaces. Largely theoretical in outlook and approach, it presents an in-depth profile of system design in scatterometers, real aperture and synthetic aperture imaging radar.

199 Ulaby, F. T., Moore, R. K. and **Fung, A. F.** *From theory to applications.* London: Adtech, 1986. 700pp. ISBN 0-201-10759-0
A very comprehensive final instalment in this trilogy, which bridges the gap between theory and practical applications of radar. Both passive and active microwave remote sensing are featured in application areas that range from sea

surface temperature and oil slick monitoring, to the backscattering properties of tree canopies and radar measurements of sea ice. The introductory chapters up-date the former volumes in this series, and the latter parts discuss current and future trends in radar remote sensing.

Oceans

200 Beal, R. C., DeLeonibus, D. C. and Katz, I. *Spaceborne synthetic aperture radar for oceanography.* Baltimore: Johns Hopkins University Press, 1981. 215pp. ISBN 0-8018-2668-3

A comprehensive technical guide to the results obtained from the short-lived Seasat ocean-dedicated satellite. This book is a valuable insight into synthetic aperture radar (SAR) that illustrates the system's great potential in many remote sensing applications. It also contains details of imagery collection over the western North Atlantic.

201 Cracknell, A. P. *Remote sensing applications in marine science and technology.* Dordrecht: D. Reidel, 1983. 466pp. ISBN 90-277-1608-0

A series of twenty papers presented at a summer school held at Dundee University that cover the general principles of remote sensing, together with its application to physical oceanography, marine resource investigations and coastal monitoring.

202 Gierloff-Emden, H. G. *Orbital remote sensing of coastal and offshore environments - a manual of interpretation.* Berlin: Walter de Gruyter, 1977. 176pp. ISBN 3-11-007278-5

Designed to provide assistance for the study, understanding, and presentation of the coastal environment through the use of satellite images and remotely sensed data. Part 2 illustrates some useful case studies, although the manual is fairly basic in its outlook and spoiled by some mistakes. References are provided at the end of each section, and there is also a list of recommended reading.

203 Houghton, J. T., Cook, A. H. and **Charnock, H.** *The study of the ocean and the land surface from satellites.* London: The Royal Society, 1983. 464pp. ISBN 0-85403-211-8

A series of papers that address geology and land use, oceans and glaciological applications. The discussions accompanying some of the sixteen papers are very informative and there are references at the end of each section.

204 Maul, G. A. *Introduction to satellite oceanography.* Dordrecht: Martinus Nijhoff, 1985. 606pp. ISBN 90-247-3096-1

A very comprehensive introduction to satellite oceanography, although the excellent content is somewhat marred by the cheap methods of printing and production. Study questions accompany chapters and reflect the book's roots in a graduate teaching programme. A bibliography, glossary and a list of symbols are included.

205 Robinson, I. S. *Satellite oceanography.* Chichester: Ellis Horwood, 1985. 455pp. ISBN 0-85312-598-8

Providing a much needed text that features an excellent general introduction to satellite oceanography and expands upon this in considerable detail with a thorough explanation of the different techniques for applying satellite data to oceanography. A discussion of this science from first principles that is suitable for postgraduates and researchers. Good for the informed and less informed reader who requires clarification of the state-of-the-art in the remote sensing of oceans.

206 Stewart, R. H. *Methods of satellite oceanography.* Berkeley: University of California Press, 1985. 360pp. ISBN 0-520-04226-3

This book describes the basic principles of electromagnetic radiation transfer from the ocean surface. While satellite techniques are emphasized, shore- and aircraft-based measurements are not ignored, resulting in a well balanced text on the subject.

207 Vernberg, F. J. and **Diemer, F. P.** *Processes in marine remote sensing.* Columbia: University of South Carolina Press, 1982. 545pp. ISBN 0-87249-411-X

A general overview of this wide field of application. A range of sensor applications are reviewed, and particular emphasis is placed on LIDAR sensing. A comprehensive list of instruments and requirements is featured.

Photogrammetry

208 Karara, H. M., ed. *Handbook of non-topographic photogrammetry.* Falls Church, Virginia: American Society of Photogrammetry, 1979. 206pp.

An excellent overview of this aspect of photogrammetry. Chapters cover terrestrial and underwater photogrammetry, holographic systems, X-ray systems and applications. Specific bibliographies are featured by topic.

209 Slama, C. C., ed. *Manual of photogrammetry.* 4th edn. Falls Church, Virginia: American Society of Photogrammetry and Remote Sensing, 1980. 1,072pp. ISBN 0-937-29401-2

An excellent and very comprehensive review of the entire field of photogrammetry, from basic theories to applications. Nineteen chapters deal with the various aspects of photogrammetry, the narrative being supported by 886 illustrations, very full bibliographies, and a section that defines the terms and symbols used in photogrammetry. Highly recommended reading for the photogrammetrist.

210 Thompson, E. H. *Photogrammetry and Surveying: a selection of papers by E. H. Thompson.* London: Photogrammetric Society, 1977. 359pp. ISBN 0-950-43761-1

A tribute to the founding father of the Photogrammetric Society. A collection of thirty-three papers that are not only of great interest, but of necessary background to the photogrammetrist. A list of publications of E. H. Thompson and editorial articles are also included.

Technical Remote Sensing

211 Chen, H. S. *Space remote sensing systems, an introduction.* Florida: Academic Press, 1985. 257pp. ISBN 0-12-170880-2

An excellent publication that gives a contemporary view of current space remote sensing systems. The basic aspects of spaceborne systems and sensors are outlined and explained with the use of technical diagrams and figures. Appendices list the observational requirements for space remote sensing and provide a list of useful acronyms. References follow each chapter.

212 Eastman Kodak Company. *Kodak data for aerial photography.* 5th edn. Rochester, New York: Eastman Kodak Company, 1982. 137 pp. ISBN 0-87985-298-4.

A concise manual that lists the aerial-photograph-related products from this company. The topics covered range from film types through processing media and filters to photographic properties. Many tables, diagrams, data sheets and a guide to other Kodak publications.

213 Egan, W. E. *Photometry and polarization in remote sensing.* New York: Elsevier Science Publishing Company, 1985. 512pp. ISBN 0-444-00892-6

A highly practical text suitable for all researchers involved in optical remote sensing. Divided into four sections, which cover Introduction to photometry and polarization, Mathematical bases, Optical fundamentals and Applications, this volume is both technical and practical in its approach to this aspect of remote sensing.

214 Holkenbrink, P. F. *Manual on characteristics of Landsat computer-compatible tapes produced by the EROS Data Center digital image processing system.* Washington D.C.: United States Geological Survey, 1978. 77pp.

A well illustrated guide to sensor system characteristics of the RBV and MSS systems carried on Landsats 1, 2 and 3. A general introduction is followed by tape specifications for both devices. Eight references are cited.

215 Hunting Geology and Geophysics Limited. *Evaluation of SPOT simulation imagery over the Derbyshire test site.* Borehamwood, Hertfordshire: Hunting Geology and Geophysics and Hunting Technical Services Limited, NRSC, 1984. 21pp.

This booklet outlines the uses of SPOT simulation imagery with regard to geomorphology, geology, agriculture, forestry and land use. Comparisons of SPOT simulation with other imagery are given. Well illustrated, but fairly lightweight as regards technical information content.

216 Killenger, D. K. and **Mooradian, A.,** eds. *Optical and laser remote sensing*. Heidelberg: Springer-Verlag, 1983. 383pp. ISBN 3-540-12170-6
This monograph features a compilation of papers which represent an overview of the present state of development in optical and laser remote sensing technology. The subjects covered include both passive and active remote sensing techniques in the ultraviolet and infrared spectral regions.

217 Measures, R. M. *Laser remote sensing, fundamentals and applications*. New York: John Wiley & Sons, 1984. 510pp. ISBN 0-471-08193-0
A comprehensive text that presents an enthusiastic review of the use of lasers in remote sensing. The history, fundamentals, systems and applications of LIDAR are discussed with a detailed and technical approach.

218 Schanda, E. *Physical fundamentals of remote sensing*. Heidelberg: Springer-Verlag, 1986. 187pp. ISBN 0-387-16236-4
A useful, in-depth and technical analysis of the basic theories upon which remote sensing is founded. Based on the author's teaching experience, this text aims to impart knowledge of the significant interactions between electromagnetic radiation and matter that occur in remote sensing.

219 Slater, P. N. *Remote sensing optics and optical systems*. Reading, Massachusetts: Addison-Wesley, 1980. 575pp. ISBN 0-201-07250-5
A foundation for applications research in remote sensing incorporating optical techniques and sensors. Practical as well as theoretical considerations are described, as are some of the aerial, and all the space remote sensing systems flown (to its publishing date), and their performance analysed. There are extensive technical appendices and references, together with abbreviation and symbol lists at the front.

220 SPOT Image Corporation. *SPOT simulation applications handbook*. Falls Church, Virginia: American Society of Photogrammetry, 1984. 274pp. ISBN 0-937294-60-8
This monograph contains a collection of thirty-two papers presented at the 1984 SPOT Symposium. Areas covered include an overview of the SPOT satellite system, geology, agriculture, urban planning and cartography, water resources, environmental applications and forestry. References follow each paper, and some high-quality prints of SPOT simulation imagery are filed at the rear.

221 Swain, P. H. and **Davis, S. M.** *Remote sensing: the quantitative approach*. New York: McGraw-Hill, 1978. 396pp. ISBN 0-07-062576-X
Primarily for user education, with problems that need to be worked out. A useful guide, but in consideration of recent developments in sensor data and processing technologies, it is rapidly becoming outdated. The theoretical and practical bases of remote sensing and how the two interact are detailed, but the book is limited by the fact that remote sensing is multidisciplinary, and all of its aspects cannot possibly be covered. However, it is a very useful text, as currently there is

nothing to replace it. References are given at the end of each chapter and a glossary of useful terms is provided, together with a list of practical exercises.

Training

222 Hamblin, W. K. *Atlas of stereoscopic aerial photographs and Landsat imagery of North America.* Minnesota: TASA Publishing Company, 1980. 102pp. ISBN 0-935698-00-0

A useful and practical teaching aid that uses ninety-one stereo pairs and twenty-two Landsat scenes that depict the details of size, shape and spatial relations of landforms and rock bodies.

223 *ITC Information Booklets* I
International Institute for Aerospace Survey and Earth Sciences, P.O. Box 6, NL-7500 AA Enschede, The Netherlands

224 ITC Publications. *Series A (Photogrammetry)* I
International Institute for Aerospace Survey and Earth Sciences, P.O. Box 6, NL-7500 AA Enschede, The Netherlands

225 ITC Publications. *Series B (Photointerpretation)* I
International Institute for Aerospace Survey and Earth Sciences, P.O. Box 6, NL-7500 AA Enschede, The Netherlands

226 Lillesand, T. M. and **Kiefer, R. W.** *Remote sensing and image interpretation.* New York: John Wiley & Sons, 1979. 612pp. ISBN 0-471-02609-3

Widely regarded as the best introductory text for use in remote sensing. The sections on airborne remote sensing systems are comprehensive, but both these, and those on satellite remote sensing, are in need of urgent updating.

227 Short, N. M. *The Landsat tutorial workbook: basics of satellite remote sensing.* Washington D.C.: NASA. 553pp.

Conceived as a self-training aid to displace costly applied remote sensing courses. This publication fulfils the role admirably with a combination of informative text and illustrations, plus a series of practical exercises at the end of each chapter. Many topics normally found in standard texts are not covered, or are treated in an unconventional manner, the emphasis being on practicality and applicability. Depending on skill and experience, the full training procedure should take approximately forty hours. One of the most complete introductions to the application of satellite remote sensing. Glossary, list of data sources, selected references.

228 Techniques rurales en Afrique. *II manuel de photointerprétation.* Paris: Secrétariat d'Etat aux Affaires Etrangères, 1970. 248pp.

This workbook is a companion to the six sets of teaching folders that accompany it. The teaching sets focus on film types, geology and geomorphology, pedology, vegetation, rural habitats and archaeology and urban areas. Stereo air photo prints are included in monochrome, colour and colour infrared, together with overlays and diagrams. The main work guide is in French, but the sets have brief explanatory notes in English on the reverse of the stereo pairs. The back of the guide features bibliographies and a glossary.

229 **Wright, J.** *Ground and air survey for field scientists.* Oxford: Clarendon Press, 1982. 327pp. ISBN 0-19-857560-2
A detailed explanation of simple field and air survey techniques is given, together with an indication of more advanced survey techniques. Map and technique analysis are covered and all the technical aspects of the title are fully detailed. Glossary of technical terms and a recommended further-reading list.

Water Resources

230 **Deutsch, M., Weisnet, D. R.** and **Rango, A.** *Satellite hydrology.* Minneapolis: American Water Resources Association, 1981. 730pp.
The proceedings of the 5th Annual William T. Pecora Memorial Symposium on Remote Sensing, which contains 120 papers ranging from topics such as hydrological rainfall estimation, to soil moisture determination. References are provided after each paper.

231 **Salomonson, V. V.** and **Bhavsar, P. D.** *The contribution of space observations to water resources management.* Oxford: Pergamon Press, 1980. 280pp. ISBN 0-08-024473-4
A collection of papers that review important research and development in the fields of hydrology and water resources. The potential of second-generation satellite systems for water resource management is considered. Citations are given at the end of each paper and an author index is included.

232 **Thompson, K. P. B., Lane, R. K.** and **Csallony, S. C.,** eds. *Remote sensing and water resources management.* Illinois: American Water Resources Association, 1973. LOC cataloguing number: 73-90430
Proceedings of a conference designed to illustrate the use of remote sensing in water resources and to encourage collaboration between remote sensing scientists and water resources researchers. A variety of interesting papers are presented, covering many forms of imagery and their relative applications.

Reports

233 *ESA Contractor Reports*
European Space Agency, 8-10 rue Mario Nikis, F-75738 Paris Cedex 15, France

234 *ESA Scientific-Technical Reports, Notes and Memoranda* I
European Space Agency, 8-10 rue Mario Nikis, F-75738 Paris Cedex 15, France

235 *Government Reports Announcements and Index* bi-W
U.S. National Technical Information Service, Springfield, VA 22161, U.S.A.
This very useful series reflects the report literature stemming from government-sponsored research, and research by government agencies, contractors and grantees. Reports are divided into twenty-two subject fields, and each citation is accompanied by an abstract.

236 *National Aeronautics and Space Administration. Technical Memorandums* I
National Technical Information Service, 5285 Port Royal Road, Springfield, VA 22151, U.S.A.

237 *National Aeronautics and Space Adminstration, Technical Notes* I
National Technical Information Service, 5285 Port Royal Road, Springfield, VA 22151, U.S.A.

238 *National Aeronautics and Space Administration, Technical Reports* I
Superintendent of Documents, Washington, DC 20402, U.S.A.

239 *NTIS Technical Notes (incorporating NASA Technical Briefs)* M
National Technical Information Service, 5285 Port Royal Road, Springfield, VA 22151, U.S.A.
One-page reviews of the current state-of-the-art in a range of U.S. Federal agency activities.

240 *Scientific and Technical Aerospace Reports* semi-M
U.S. National Aeronautics and Space Administration, Scientific and Technical Information Facility, Box 8757, Baltimore-Washington International Airport, MD 21240, U.S.A.
The literature scanned for this indexing and abstracting periodical includes patents, dissertations and translations, together with reports from academic and commercial institutes and organizations, and government agencies and their contractors. A high percentage of the citations are on remote sensing and related disciplines, and the impact of this periodical is demonstrated by the amount of remote sensing abstracting and indexing publications which name it as a source.

Conference Material

241 *Conference Papers Index* M
Cambridge Scientific Abstracts, 5161 River Road, Bethesda, MD 20816, U.S.A.
A very up-to-date record of conference proceedings that uses final programmes and journals as its information sources. The meeting title, location, sponsors, and list

of papers are included, but information on where to obtain the printed proceedings is often not available at the time that this publication's entries are compiled.

242 *Ei Engineering Conference Index* A
(Six hardcover books - plus a two-volume cumulative index)
Engineering Information Inc., 345 East 47th Street, New York, NY 10017, U.S.A.

This useful guide provides rapid and convenient access to the proceedings of more than 1,600 worldwide conferences in engineering and related disciplines. The conference literature of engineering is divided into six different areas (several of which are of interest to those involved in remote sensing) and a two-volume cumulative index aids in the location of cross-referenced material.

243 *Index and Abstracts 1962-1980*
Enviromental Research Institute of Michigan, P.O. Box 8618, Ann Arbor, MI 48107, U.S.A.

Covers twenty-eight printed proceedings volumes.

244 *Index of Conference Proceedings Received* M
British Library Document Supply Centre, Boston Spa, Wetherby, West Yorkshire LS23 7BQ, England

This publication has been available since 1965 and includes records of the conference literature in all subject fields.

245 *Index to Scientific and Technical Proceedings (ISTP index)* M
Institute for Scientific Information, 3501 Market Street, Philadelphia, PA 19104, U.S.A.

This publication attempts to cover 'significant' conference proceedings, and the individual papers' titles for these meetings are included, together with ancilliary information on conference sponsors, title, date, place, how much the proceedings cost, and from where they may be obtained.

246 *Index to Technical Papers from Annual and Fall Meetings of ASPRS & ACSM 1975-1982 & 1983*
American Society for Photogrammetry and Remote Sensing, 210 Little Falls Street, Falls Church, VA 22046-4398, U.S.A.

A comprehensive index to thirty-three conference proceeding volumes.

247 *The ISPRS: Organization and Programs 1984-1988*
International Society for Photogrammetry and Remote Sensing (ISPRS), Department of Photogrammetry, S-100 44 Stockholm, KTH, Sweden

An overview of the structure and function of the ISPRS.

248 *World Meetings: Outside the United States and Canada* Q
Macmillan Publishing Company, Professional Books Division, 886 Third Avenue, New York, NY 10022, U.S.A.

249 *World Meetings: United States and Canada* Q
Macmillan Publishing Company, Professional Books Division, 886 Third Avenue, New York, NY 10022, U.S.A.

Trade Literature

250 *Construction and Civil Engineering Index* Q
Technical Indexes Limited, Willoughby Road, Bracknell, Berkshire RG12 4DW, England

251 *Electronic Engineering Index* 3 p.a.
Technical Indexes Limited, Willoughby Road, Bracknell, Berkshire RG12 4DW, England

252 *Sensing the Earth* I
VEB Carl Zeiss Jena, Carl-Zeiss-Strasse 1, 6900 Jena, German Democratic Republic
A well illustrated guide to the equipment available from Carl Zeiss Jena.

Literature Guides and Bibliographies

253 *Bibliography and Index of Geology* M
American Geological Institute, 4220 King Street, Alexandra, VA 22302, U.S.A.
The citations in this bibliography reflect the content of their parent host, the GEOREF database. A great many of the entries on remote sensing and aerial photography in geology are found under the Geophysics subheading.

254 *Bibliography of Remote Sensing Publications* I
Remote Sensing Laboratory, Center for Research, University of Kansas, 2291 Irving Hill Drive, Lawrence, KA 66045, U.S.A.
A supplement to the publications list of the Remote Sensing Laboratory, one of the most active university groups in the U.S.A.

255 *Books in Print* Q
R. R. Bowker Company, Database Publishing Group, 205 E. 42nd Street, New York, NY 10017, U.S.A.

256 *The Bowker International Serials Database Update* Q
R. R. Bowker, 254 W. 17th Street, New York, NY 10011, U.S.A.
This publication contains all of the information necessary to remain up to date between successive editions of other Ulrich periodical publications.

257 *British National Bibliography* W
British Library Bibliographic Services Division, 2 Sheraton Street, London W1V 4BH, England
As the main bibliography for Great Britain, the *British National Bibliography* aims to list and provide a brief abstract of every new work (including journals) published within the U.K.

258 Bryan, M. L. *Remote sensing of earth resources: a guide to information sources.* Detroit: Gale Research Company, 1979. 188pp. ISBN 0-8103-1413-4
A very well-laid-out book that makes extensive use of cross referencing between its 378 annotated items. Most of the descriptions of entries are brief, and the book does suffer slightly from a lack of narrative; however, considering its age it is still a remarkably pertinent reference source, although several sections are well out of date.

259 Carter, D. J. *The remote sensing sourcebook: a guide to remote sensing products, services, facilities, publications and other materials.* London: McCarta, 1986. 175pp. ISBN 0-906318-15-7
While this book is a useful guide to the items in its title, it is unfortunately let down by a poor layout that tends to obscure the substantial amount of information contained; this problem is compounded by a scanty index and lack of accurate cross-referencing. These things aside, the volume is quite informative and fairly comprehensive, especially on British products, although those from North America and other countries in Europe are not so well documented.

260 *Catalogue of Official British Publications not Published by HMSO* bi-M
Chadwyck-Healey Ltd., Cambridge Place, Cambridge CB2 1NR, England

261 *Cumulative Book Index* M
H. W. Wilson Company, 950 University Avenue, Bronx, NY 10452, U.S.A.
A comprehensive listing of books published in the English language worldwide. Grey and occasional literature is largely ignored, and the entries are arranged according to the Library of Congress Filing Rules.

262 *Current Information Received* M
National Remote Sensing Centre (NRSC), R 190 Building, Space Department, Royal Aircraft Establishment, Farnborough, Hampshire GU14 6TD, England
A well presented list of publications, journals, conference literature, newsletters and books, received and stored at the NRSC library.

263 *Current Serials Received* I
British Library Document Supply Division, Publications Section, Boston Spa, Wetherby, West Yorkshire LS23 7BQ, England

264 *Daily List of Government Publications from HMSO* D
Subscriptions Department, P.O. Box 276, London SW8 5DT, England

265 *Earth Resources: A Continuing Bibliography with Indexes. NASA*
U.S. National Aeronautics and Space Administration, Washington, DC 20546, U.S.A.

This excellent bibliography contains a selection of annotated references from *Scientific and Technical Aerospace Reports* and *International Aerospace Abstracts*. The contents are divided into nine different areas: Agriculture and Forestry, Environmental Changes and Cultural Resources, Geology and Cartography, Geology and Mineral Resources, Oceanography and Mineral Resources, Hydrology and Water Management, Data Processing and Distribution Systems, Instrumentation and Sensors, and General. Several hundred abstracts are provided in each edition, and these are indexed by subject, personal author, corporate source, foreign technology, contract number, report/accession number, and accession number.

266 *HMSO Monthly Catalogue* M
HMSO Publications Centre, P.O. Box 276, London SW8 5DT, England

267 *Irregular Serials and Annuals: an International Directory* A
R. R. Bowker, 254 W. 17th Street, New York, NY 10011, U.S.A.

Nearly 36,000 entries on irregular publications, which is very useful for remote sensing, as many of its serial issues are infrequent.

268 **Krumpe, P. F.** *The World Remote Sensing Bibliographic Index.* Merrifield, Virginia: Tensor Industries Inc., 1976. 619pp.

This volume arranges over 4,000 citations on the remote sensing of natural and agricultural resources throughout the world into fourteen major subject headings and nearly fifty geographic regions. No abstracts are featured, which is a pity, but they cannot realistically be expected from a book of this size. The coverage extends from 1970 to 1976, and while this makes most of these citations dated, they may nevertheless be useful retrospectively.

269 *National Union Catalog* M
U.S. Library of Congress, Catalog Management and Publication Division, Washington, DC 20540, U.S.A.

270 (a) *Remote sensing of earth resources: a literature survey with indexes. NASA SP-7036. 1970.* 1,221pp. U.S. National Aeronautics and Space Administration, Washington, DC 20546, U.S.A.

A list of over 3,500 reports and papers that were added to the NASA Scientific and Technical Information System between 1962 and 1970. A very wide range of application areas are covered and most entries are briefly annotated. A very useful retrospective source of information that is indexed by subject, author, contract number and corporate source.

(b) *Remote sensing of earth resources: a literature survey with indexes. NASA SP-7036(01). 1970-1973 Supplement.* 654pp. U.S. National Aeronautics and Space Administration, Washington, DC 20546, U.S.A.

This two-volume bibliography continues the previous edition of the *NASA SP-7036*, and lists 4,930 documents of various types. The first volume houses the abstracts, whilst the second indexes that material by subject, author, source, contract number and report number.

271 *RESENA Bibliographies* I
Remote Sensing and Nature (RESENA) Group, 1530 Wolf Run, College Station, TX 88740, U.S.A.

Bibliographies on certain aspects of remote sensing.

272 *Ulrich's International Periodicals Directory* A
R. R. Bowker, 254 W. 17th Street, New York, NY 10011, U.S.A.

A comprehensive listing of currently available periodicals that contains over 68,000 entries from nearly 200 countries.

273 *Vance Bibliographies*
Vance Bibliographies, P.O. Box 229, 112 North Charter Street, Monticello, IL 61856, U.S.A.

274 **Wentink, N. J.** *The ITC International Bibliography of Photogrammetry (IBP).* 4th edn. Enschede: ITC, 1977. 99pp.

This dated, although useful, bibliography presents a descriptive history of the applications and theory of aerial photography, aerial surveying and photogrammetry.

Abstracts and Indexes

275 *Abstract Newsletter: NASA Earth Resources Survey Program* W
U.S. National Technical Information Service, 5285 Port Royal Road, Springfield, VA 22151, U.S.A.

Several abstracts and citations are included in each issue of this informative bulletin, together with news of reports from Federal agencies, their contractors and grantees. An annual index is produced which lists the abstracts by subject. Apart from the abstract a typical entry will include data on title, author, date, organization of pages, NTIS number, price and report number.

276 *AESIS Quarterly. (Australian Earth Sciences Information System)* Q
Australian Mineral Foundation, Private Bag 97, Glenside, S.A. 5065, Australia

An informative up-date service which abstracts and indexes material from a wide range of publications and documents, including a very large number of reports, many of which are derived from the mining and related industries. Remote sensing and cartography are featured as subject headings, and the information in this publication may be accessed online as AESIS, via the AUSINET system. Cumulative indexes for the preceding year of this quarterly current awareness bulletin are included in the December issue, and an annual microfiche cumulation is also available.

277 *AESIS Special List No. 12: Remote Sensing and Photogeology, 1976 - June 1982*
Australian Mineral Foundation, 63 Conygham Street, Private Bag 97, Glenside, South Australia 5065, Australia
A consolidated bibliographic review of this particular field that has 271 entries. The citations are grouped under seventy broad subject headings, and each one usually has between ten and twenty descriptive keywords. The list is entirely machine-generated and contains no fewer than six indexes, which are divided into subject, locality name, map reference, author, mine/deposit and stratigraphic subdivisions.

278 *Applied Science and Technology Index* M (except July)
H. W. Wilson Company, 950 University Avenue, Bronx, NY 10452, U.S.A.
Although the term remote sensing is not included in this index, references to the subject and its related disciplines may be found under the headings of Photogrammetry, Mapping - aerial, and Surveying - aerial.

279 *Computer and Control Abstracts: Science Abstracts, Section C* M
INSPEC, IEE, Station House, Nightingale Road, Hitchin, Hertfordshire SG5 1RJ, England
Generated from the INSPEC database and containing nearly 3,000 abstracts per issue, this indexing serial abstracts material from over 4,000 journals, books and proceedings each year. Remote sensing is often included (dependent on the focus of a particular paper) and there are frequent articles on related computerized themes.

280 *Current Contents: Physical, Chemical and Earth Sciences* W
Institute for Scientific Information, 3501 Market Street, Philadelphia, PA 19104, U.S.A.
A most useful current awareness service that has been available since 1961, and which reproduces and indexes the contents pages from journals and multi-authored books in a variety of scientific disciplines.

281 *Earth Science and Related Information: Selected Annotated Titles* M
Australian Mineral Foundation, Private Bag 97, Glenside, S.A. 5065, Australia
This publication contains a selective review of all of the information from Australian and international publications that is received by the Australian Mineral Foundation (AMF). Unless the title is self-explanatory each item features a brief abstract, and the journal source lists are published every January. The bulletin, which is cumulated every six months on microfiche, is computer based and has

seven indexes and a wide variety of subject headings that include aerial surveys, cartography, earth sciences information, geophysics, and remote sensing.

282 *Electrical and Electronics Abstracts: Science Abstracts, Section B* M
INSPEC, IEE, Station House, Nightingale Road, Hitchin, Hertfordshire SG5 1RJ, England
Remote sensing is included in this abstract series as and when it is related to electronic and electrical systems or applications thereof. Most of the papers mentioned deal directly with sensor technology, although there are more general entries on occasion.

283 *Engineering Index Monthly* M
Engineering Information Inc., 345 E. 47th Street, New York, NY 10017, U.S.A.
The *Engineering Index Monthly* contains abstracts and bibliographic entries that cover the world's technological literature in a large number of engineering disciplines. A high number of remote sensing citations are included in this publication, which may also be searched online as COMPENDEX. An aid to searching this bibliographic publication is the *SHE: Subject headings for engineering*, which helps in the identification of headings and titles of interest to those involved in remote sensing.

284 *Engineering Information. Technical bulletin: Remote Sensing* (discontinued 1986)
Engineering Information Inc., 345 E. 47th Street, New York, NY 10017, U.S.A.
Although this publication has now been discontinued, it is still worthwhile consulting it, especially for an introduction to specific remote sensing subjects. The categories covered range from civil and environmental engineering to electrical and electronic engineering, and nearly 1,000 very fully abstracted citations are provided.

285 *Engineering Information. Technical Bulletin: Satellites*
Engineering Information Inc., 345 E. 47th Street, New York, NY 10017, U.S.A.
This guide provides selected citations with full abstracts on the use of satellites and satellite data in a range of technologies, and in particular, engineering disciplines.

286 *Geographical Abstracts. Annual Index* A
Geo Abstracts Limited, Regency House, 34 Duke Street, Norwich NR3 3AP, England

287 *Geographical Abstracts G: Remote Sensing, Photogrammetry and Cartography* bi-M

Geo Abstracts Limited, Regency House, 34 Duke Street, Norwich NR3 3AP, England

This very comprehensive abstracting service is perhaps the best currently available for remote sensing. A very wide range of subjects and applications are covered under the general headings of remote sensing, photogrammetry and orthophotography, surveying, cartography, data acquisition, interpretation and applications, history of surveying and cartography, and atlases, sheet maps and librarianship. Over 3,000 abstracts are provided each year in the above disciplines, and the information may also be retrieved online as the GEOBASE file.

288 *Index to Scientific Book Contents (ISBC)* Q

Institute for Scientific Information, 3501 Market Street, Philadelphia, PA 19104, U.S.A.

This useful index acts as a guide to the contents of multi-authored publications and the series is particularly applicable to remote sensing, because of the numbers of published conference proceedings that dominate certain areas of the science. Often a title alone may be either misleading or insufficient indication of content, but use of this publication reveals individual authors and chapters in the case of books, and individual authors and papers in the case of conference proceedings.

289 *International Aerospace Abstracts* semi-M

American Institute of Aeronautics and Astronautics, Technical Information Service, 555 W. 57th Street, New York, N.Y. 10019, U.S.A.

A broad coverage of the aerospace sciences is catered for by this serial, which includes a wide variety, and large number, of citations specific to remote sensing.

290 *Monthly Catalog of United States Government Publications* M

U.S. Government Printing Office, Washington, DC 20402, U.S.A.

A regular bulletin on the U.S. government's publishing activities that includes remote sensing and a large number of references to maps produced by the U.S. Geological Survey.

291 *Pascal Explore: Part 48: Environment Cosmique Terrestre, Astronomie et Géologie Extraterrestre* 10 p.a.

Centre National de la Recherche Scientifique, Centre de Documentation Scientifique et Technique, Service des Abonnements, 26 rue Boyer, F-75971 Paris 20, France

292 *Pascal Explore: Part 49: Météorologie* 10 p.a.

Centre National de la Recherche Scientifique, Centre de Documentation Scientifique et Technique, Service des Abonnements, 26 rue Boyer, F-75971 Paris 20, France

293 *Physics Abstracts: Science Abstracts, Section A* bi-M
INSPEC, IEE, Station House, Nightingale Road, Hitchin, Hertfordshire SG5 1RJ, England
This journal's coverage embraces all of the general aspects of this science, as well as several more specialized areas. The inclusion of remote sensing is dependent upon what physical aspects are under consideration, and many of the entries fall under fields such as atmospheric phenomenology, and electrical, magnetic and optical properties of the atmosphere.

294 *Referativnyi Zhurnal Geodeziya i Aeros'emka* M
Vsesoyuzny Institut Nauchno-Tekhnicheskoi Informatsii (VINITI), Baltiiskaya ul. 14, Moscow A-219, Russian S.F.S.R., U.S.S.R.
Over 3,000 abstracts are published annually in this journal, all in the Russian language. Although the entries concentrate on geodesy, aerial photography, aerial surveying and photogrammetry are also quite strongly represented.

295 *Science Citation Index* W
Institute for Scientific Information, University City Science Center, Philadelphia, PA 19104, U.S.A.
A massive and exhaustively comprehensive calendar year index that provides an in-depth coverage of all subjects. Remote sensing and related disciplines are very well documented and this publication is a useful means of elaborating on preliminary results from a literature search.

State-of-the-art Reviews

296 *Annual Review of Earth and Planetary Sciences*
Annual Reviews Inc., 4139 El Camino Way, Palo Alto, CA 94306, U.S.A.

297 *Digital image processing: a critical review of technology (SPIE Vol. 528).* 1985. 26 pp.
The International Society for Optical Engineering, P.O. Box 10, Bellingham, WA 98227, U.S.A.
This volume provides a very thorough overview of algorithms and concepts, implementation issues, application fields and miscellaneous applications.

298 *Index to Scientific Reviews* semi-A
Institute for Scientific Information, 3501 Market Street, Philadelphia, PA 19104, U.S.A.
A very comprehensive publication that contains reviews of subjects in more than 100 disciplines from over 3,000 journals and periodic review series.

299 *Recent advances in civil space remote sensing (SPIE Vol. 481).* 1984. 273pp.

The International Society for Optical Engineering, P.O. Box 10, Bellingham, WA 98227, U.S.A.

A comprehensive look at remote sensing applications.

300 *Remote sensing: a critical review of technology (SPIE Vol. 475).* 1984. 159pp.

The International Society for Optical Engineering, P.O. Box 10, Bellingham, WA 98227, U.S.A.

The proceedings of a conference that covered atmospheric effects, marine and geological research, spectral signature studies, agriculture, information processing, engineering, mapping, and future prospects for remote sensing.

301 *Remote Sensing of Natural Resources: A Quarterly Literature Review* Q

University of New Mexico, Technology Application Center, Albuquerque, NM 87131, U.S.A.

302 *Remote Sensing Reviews* I

Harwood Academic Publishers, 50 W. 23rd Street, New York, NY 10010, U.S.A.

The major review serial for remote sensing. Each edition provides in-depth coverage of a wide range of both practical and theoretical topics in remote sensing. Subject indexes are included, together with very comprehensive lists of references.

Online Databases

303 *Directory of Online Databases* Q

Cuadra-Elsevier Partners, 2001 Wiltshire Blvd., Ste. 305, Santa Monica, CA 90403, U.S.A.

The most up-to-date directory available of information concerning online databases. Very useful for checking the current status of, and developments in, the computerization of information on remote sensing and related subjects.

304 *Published Search Catalog* semi-A

National Technical Information Service, U.S. Department of Commerce, Springfield, VA 22161, U.S.A.

This catalogue presents the coverage and topic of well-structured online searches from the NTIS database in around 3,000 subject areas. The current (1986) edition includes approximately twenty-five entries from remote sensing, photogrammetry, pattern recognition and image processing. These publications provide rapid and economical means of keeping up to date with a particular area, and are very useful for those groups or individuals who wish to utilize online information, but do not have their own in-house access to such services.

305 *RESORS Keyword Dictionary*

Canada Centre for Remote Sensing, Department of Energy, Mines and Resources, 2464 Sheffield Road, Ottawa, Ontario K1A 0Y7, Canada
An innovative and very comprehensive glossary of search terms for the RESORS database that may also be of use when selecting search descriptors for other online databases.

306 Williams, M. E., ed. *Computer readable databases: A Directory and Data Sourcebook.* Chicago: American Library Association, 1985
Over 5,000 databases are exhaustively listed in this two-volume guide, of which Volume 1, *Science, Technology and Medicine*, is the more relevant to remote sensing.

307 AEROSPACE DATABASE
Origin: American Institute of Aeronautics and Astronautics, Technical Information Service and the National Aeronautics and Space Administration, Scientific and Technical Information Branch
Printed equivalent: International Aerospace Abstracts, Scientific and Technical Aerospace Reports
Available from: 1962
Approx. File size: 1,600,000 items
File update: 6,000 items a month
User guides: Online thesaurus which is also available in hardcopy as the NASA Thesaurus
Searchable through: DIALOG; MDC
Information: Access to this bibliographic reference file is restricted; it is available as the NASA file to member states of the European Space Agency through ESA-IRS. The database concentrates on aeronautics, astronautics, space sciences and the geosciences. It is particularly useful for satellite technology and remote sensing

308 AGRIS
Origin: Food and Agriculture Organization, AGRIS Coordinating Centre
Printed equivalent: *AGRINDEX*
Available From: 1975
Approx. File size: 1,200,000 items
File update: 10,000 items each month
User guides: AGROVO Multilingual Macrothesaurus, AGRIS: STDIRS User Manual
Searchable through: DIMDI; ESA-IRS
Information: AGRIS provides access to agricultural science and technology literature. Remote sensing records are available, some with abstracts

309 AESIS (Australian Earth Sciences Information System)
Origin: Australian Mineral Foundation
Printed equivalent: AESIS Quarterly

Available from: 1975
Approx. File size: 30,000 items
File update: 700 items per quarter
User guides: AESIS guide is available on AUSINET
Searchable through: AUSINET
Information: The AESIS file lists a wide range of geosciences and remote sensing documentation relating to continental Australia. Over ninety percent of the records are from open-file company reports and periodicals. Map sheet references are given in addition to the normal bibliographic information

310 COMPENDEX
Origin: Engineering Information Inc.
Printed equivalent: *Engineering Index, Bioengineering Abstracts, Energy Abstracts*
Available from: 1969
Approx. File size: 1,500,000 items
File update: 10,000 items a month
User guides: *Publications Indexed for Engineering (PIE)*, a list of journal and conference publications scanned by Engineering Information, Inc.; *Subject Headings for Engineering (SHE)*, a classification and indexing tool to aid searching the database. Various other user aids and guides
Searchable through: BRS; CAN/OLE; CEDOCAR; CISTI; DATA-STAR; DIALOG; ESA-IRS; FIZ Technik; INKA; Pergamon InfoLine; SDC
Information: Covers all engineering disciplines

311 GEOBASE
Origin: Geo Abstracts Ltd.
Printed equivalent: *Geographical Abstracts. Part G: Remote Sensing, Photogrammetry and Cartography*
Available from: 1980
Approx. File size: 200,000 items
File update: 3,000 records each month
User guides: -
Searchable through: Dialog
Information: This file covers the worldwide literature on a number of subjects, including remote sensing and related disciplines. Over 5,000 journals are scanned for this database, together with more than 2,000 books, monographs, conference proceedings and reports

312 GEOLINE
Origin: Federal Institute for Geosciences and Natural Resources
Printed equivalent: None
Available from: 1969
Approx. File size: 400,000 items
File update: 2,000 items each month
User guides: None

Searchable through: INKA
Information: GEOLINE covers geoscience literature from journals, monographs and government reports. Surveying, mapping, photogrammetry and remote sensing are all included as subjects and the records are published in English, French, German, Russian and several other languages

313 GEOREF (Geological Reference File)
Origin: American Geological Institute
Printed equivalent: *Bibliography and Index of Geology*
Available from: 1961
Approx. File size: 950,000
File update: 5,000 items each month
User guides: *GeoRef Thesaurus and Guide to Indexing, GeoRef Serials and KWOC Index*
Searchable through: CIRTI; DIALOG; SDC
Information: This database provides worldwide coverage of the geosciences literature, and the geophysics and hydrology sections feature remote sensing citations

314 IEEE FINDING YOUR WAY
Origin: Institute of Electrical and Electronics Engineers Inc.
Printed equivalent: Various IEEE publications
Available from: 1985
Searchable through: IEEE
Information: This database has two particularly interesting files, the Tutorial Database file and Catalog file, which respectively provide information on forthcoming meeings and courses, and full bibliographic records of all IEEE publications

315 INSPEC
Origin: Institution of Electrical Engineers
Printed equivalent: *Physics Abstracts (PA), Electrical and Electronic Abstracts (EEA), Computer and Control Abstracts (CCA)*
Available from: 1969
Approx. File size: 2,300,000 items
File update: 17,000 items each month
User guides: INSPEC Thesaurus, INSPEC User Manual, List of Journals and Other Serial Sources
Searchable through: BRS; CAN/OLE; CEDOCAR; DATA-STAR; DIALOG; ESA-IRS; INKA; JICST; SDC; Tskuba University
Information: The published *Science Abstract* series are generated from this database and they all contain discipline-specific citations on remote sensing. Details from all areas of computing, control, engineering, electronics, physics and information technology are included and copies of the documents cited may be ordered online via the terminal's PrimorDial service

316 ISPRS Information Retrieval System (ISPRS-IRS)
Origin: International Society for Photogrammetry and Remote Sensing (ISPRS)
Printed equivalent: A range of abstracting and indexing serials are being considered
Available from: -
Approx. File size: -
File update: An estimated 5,000 items annually
User guides: -
Searchable through: A major host, possibly ESA-IRS
Information: It is the aim of the ISPRS to implement an Information Retrieval System. The database will offer bibliographic records in the three official ISPRS languages, English, French and German. It is expected that co-operation in database compilation will come from several other organizations that offer bibliographic services. The range of subject fields will include remote sensing, photogrammetry, cartography and surveying

317 LEDA
Origin: ESA Earthnet Programme Office
Sensor type: MSS from all satellites, RBV from Landsat-3, TM on Landsats 4 and 5. EROS Data Centre Catalog; MSS from Landsats 1 to 5, TM data from Landsats 4 and 5
Available from: 1972 (EROS Data Centre), 1975 (Fucino), 1978 (Kiruna), 1984 (Maspalomas)
Approx. File size: 800,000 items
File update: 1,500 items each week
User guides: LEDA User Manual
Searchable through: ESA-IRS
Information: This database catalogues the imagery archived from Landsats 1, 2, 3, 4 and 5 available at ESA/Earthnet/Eurimage following acquisition at Fucino. The EROS Data Center Catalog, which has worldwide coverage, is also included in the database. Geographical location information related to the centre of each acquired scene may be determined through latitude/longitude coordinates or via the World Reference System of Path/Row numbers. Further data concern the satellite used, sensors, quadrant cloud coverage, image quality, sun azimuth/elevation and acquisition information

318 NTIS (National Technical Information Service Bibliographic Database)
Origin: National Technical Information Service
Printed equivalent: *U.S. Government Reports Announcements and Index*
Available from: 1964
Approx. File size: 1,200,000 items
File update: 5,000 items each month
User guides: None
Searchable through: BRS; DATA-STAR; DIALOG; ESA-IRS; FIZ-TECHNIK; INKA; MDC; SDC

Information: The NTIS database records research, development and engineering reports from U.S. Federal government agencies, their contractors and grantees. Sponsored translations and selected abstracts from the Aerospace (NASA) database, *Scientific and Technical Aerospace Reports*, and the National Standards Association are also included

319 PASCAL
Origin: Centre de Documentation Scientifique et Technique du Centre National de la Recherche Scientifique
Printed equivalent: *PASCAL Folio, PASCAL Explore*
Available from: 1973
Approx. File size: 6,000,000 items
File update: 35,000 items each month
User guides: *Thesaurus, Manual d'utilisation PASCAL, Catalogue CDST, Plan de Classement PASCAL*
Searchable through: BNDO-IFREMER; ESA-IRS; TELESYSTEMES-QUESTEL
Information: The main PASCAL M database may be searched in the form of separate, subject-specific files that correspond to the printed equivalents. The sciences and technology are very well covered and this database can be particularly useful for retrieval of electrical engineering and computer science literature. Nearly ninety-five percent of the file input is from journals and there is also coverage of reports, proceedings, patents and theses

320 RESORS (Remote Sensing Online Retrieval System)
Origin: Canada Centre for Remote Sensing
Printed equivalent: RESORS Microfiche
Available from: 1900
Approx. File size: 55,000 items
File update: 500 items each month
User guides: *RESORS Keyword Dictionary, RESORS Users Guide, Slide Dictionary*
Searchable through: DATAPAC; RESORS (via letter, telex or telephone)
Information: RESORS provides access to a wide range of bibliographic references concerned with the techniques, instrumentation and applications of remote sensing, photogrammetry and digital image processing. Reports, journals and conference literature are available in English and French. RESORS Slide Collection is an accompanying slide database, accessible through the RESORS search program, that contains over 5,000 slides which may be used in conjunction with the technical papers found in the RESORS document database. Some of the slide sets have an accompanying cassette tape or brochure. Lists of present, past and future satellite systems and acronyms are available as non-computer-readable information

321 SATELDATA
Origin: European Space Research and Technology Centre (ESTEC)
Printed equivalent: None

Available from: 1974
Approx. File size: 1,000 items
File update: 100 items annually
User guides: None
Searchable through: ESA-IRS
Information: The information stored in SATELDATA is derived from satellite hardware equipment reports. The technical details and specifications of design and performance are given. Data are available on both current and future systems including HELIOS, GEOS, METEOSAT, ECS/MARECS, HIPPARCOS and L-SAT

322 SATELLITE NEWS
Origin: Philips Publishing Inc.
Printed equivalent: *Satellite News*
Available from: 1982
Approx. File size: Current
File update: Weekly
User guides: None
Searchable through: NewsNet
Information: A current full-text aerospace bulletin, which covers policies and developments, research, commercialization and news

323 SIGLE (System for Information on Grey Literature in Europe)
Origin: Various European documentation centres
Printed equivalent: Various
Available from: 1981
Approx. File size: 60,000 items
File update: 3,000 items each month
User guides: None
Searchable through: BLAISE; INKA
Information: SIGLE covers 'grey' literature, i.e. documents such as informal research reports, proceedings, theses, notes, private communications and any material that is not regularly published or subject to classification procedures

324 SPACE COMMERCE ONLINE
Origin: Space Information Online
Printed equivalent: None
Available from: 1985
Approx. File size: 1,000,000
File update: Variable
User guides: None
Searchable through: Producer
Information: Particularly useful for retrieving bibliographic citations on the commercialization of remote sensing, as well as information on aerospace policies, legislation and news

325 SPACE COMMERCE BULLETIN
Origin: Television Digest Inc.
Printed equivalent: *Space Commerce Bulletin*
Available from: 1984
Approx. File size: 600 pages
File update: 300 pages every two months
User guides: None
Searchable through: NewsNet
Information: Space Commerce Bulletin is a full-text file concerned with the commercial development of space. Space technology, politics, economics and the commercialization of remote sensing are amongst the topics covered

326 SPACECOMPS
Origin: European Space Agency
Printed equivalent: None
Available from: 1971 (reports)
Approx. File size: 12,000 items
File update: 500 items annually
Searchable through: ESA-IRS
Information: A full, very technical and comprehensive catalogue of spacecraft components and associated performance reports. The reports cited in this database date back until 1970, while the component lists are current

327 SPACE PATENTS
Origin: European Space Agency Commercialization Office
Printed equivalent: None
Available from: Current file
Approx. File size: 10,000 items
File update: 500 items annually
User guides: Online tutorials
Searchable through: ESA-IRS
Information: The Space Patents file contains current references to space-technology-related patents. Copies of patents cited on the database are available from the respective National Patent Offices of the country of interest

328 SPACESOFT
Origin: Computer Software Mangement and Information Center, Georgia University
Printed equivalent: *International Catalog of Computer Programs*
Available from: 1983
Approx. File size: 1,200 items
File update: Monthly
User guides: None
Searchable through: ESA-IRS, DIALOG

Information: This file lists publicly available computer programs that have been produced by NASA, Federal agencies, commercial companies and academic institutions. The coverage is broad, encompassing aeronautics and astronautics, engineering, geosciences, mathematical and computer sciences and space sciences

329 TECHNICAL REFERENCE FILE
Origin: National Space Science Data Center
Printed equivalent: None
Available from: 1971
Approx. File size: 36,000 items (12 records per item)
File update: Weekly
User guides: None
Searchable through: MODCOMP
Information: This file is a sub-set of the National Space Science Data Center Automated Bibliographic Reference File, and contains information from reports and other documents concerned with spacecraft instrumentation

Newsletters and Bulletins

330 *ACSM News* I
American Congress on Surveying and Mapping, 210 Little Falls Street, Falls Church, VA 22046, U.S.A.
A newsbrief type of publication that reports on society events and research.

331 *Alberta Remote Sensing Centre* A
Alberta Remote Sensing Centre, 11th floor, 9820 106 Street, Edmonton T5K 2J6, Alberta, Canada

332 *ARRSTC Newsletter* Q
Asian Regional Remote Sensing Training Centre (ARRSTC), Asian Institute of Technology (AIT), P.O. Box 2754, Bangkok 10501, Thailand
An informative quarterly publication that gives news of students, project up-dates, progress reports, maps and surveys of work in Asia. There are useful sections that deal with conferences, workshops, seminars, book reviews and news briefs.

333 *Asian Association on Remote Sensing* Q
Asian Association on Remote Sensing Secretariat (AARS), Institute of Industrial Science, University of Tokyo, 7-32 Roppongi, Minatoku, Tokyo, Japan
Published since 1983, this newsletter is distributed to member countries of the Asian Association on Remote Sensing and to interested non-member countries and organizations. The coverage includes minutes of conferences, regional activities and technical and academic information on remote sensing.

334 *Australian Landsat Station Newsletter* bi-M

Australian Landsat Station (ALS), 22-36 Oatley Court, P.O. Box 28, Belconnen, ACT 2616, Australia

A glossy publication that contains a calendar of Landsat flights across Australia, ALS news, details of the Australian remote sensing community, reports on various sensors, product descriptions and international news. There are also short technical papers on satellite imagery interpretation, and lists of the Landsat products that are available from ALS, advance conference warnings, and a very useful index list of ALS reference centres.

335 *Boletim da Produçao Téchnico-Científica du IPE* M

Instituto de Pesquisas Espaciais, P.O. Box 515, São José dos Campos - SP, São Paulo State 12225, Brazil

A Portuguese-language publication that includes abstracts of papers.

336 *Boletín Informativo del Instituto Geográfico Militar* I

Instituto Geográfico Militar, Sección Sensores Remotos, Nueva Santa Isabel No. 1640, Santiago, Región Metropolitana, Chile

337 *British Pattern Recognition Association Newsletter* I

Department of Electronic and Electrical Engineering, University of Surrey, Guildford GU2 5XH, England

News, advance conference warning, a diary and announcements are featured in this publication.

338 *Canadian Advisory Council in Remote Sensing (Annual minutes)*

Canada Centre for Remote Sensing, Department of Energy, Mines and Resources, 2464 Sheffield Road, Ottawa K1A 0Y7, Canada

Published in English and French, this useful annual summarizes the status of Canadian remote sensing programmes, policies and research.

339 *Centro de Levantamientos Intergrados de Recursors Naturales por Sensores Remotos Newsletter* I

Centro de Levantamientos Intergrados de Recursors Naturales por Sensores Remotos, Calle Paz y Mino s/n Edf. I.G.M 4 to. piso/Casilla 8216, Quito, Ecuador

340 *CESTA SELF* bi - M

S.E.L.F., 47 bis rue du Rocher, F-75008 Paris, France

Brief news reports in French.

341 *The Clears Review* Q

Cornell Laboratory for Environmental Application of Remote Sensing, Cornell University, 464 Hollister Hall, Ithaca, NY 14853, U.S.A.

This quarterly newsbrief contains details of activities, courses and conferences at

Cornell. A fairly extensive, and very useful section lists selected articles and other publications from a wide variety of sources.

342 *Columbus Logbook* Q
ESA Publications Division, ESTEC, Postbus 299, NL-2000 AG Noordvijk, The Netherlands
Contains information on the policies, programmes and events that affect ESA's In-orbit infrastructure programme.

343 *Daedalus International* I
Daedalus International Inc., P.O. Box 1869, Ann Arbor, MI 48106, U.S.A.
A publicity mailshot that advertises Daedalus products and gives notice of forthcoming conferences.

344 *Dornier ERS 1* Q
Dornier System GmbH, Department VR, P.O. Box 1360, D-7990 Friedrichshafen 1, Federal Republic of Germany
Highlights the progress made on ERS-1 and reviews its potential applications in remote sensing.

345 *DSIR Newsletter* A
Division of Information Technology, DSIR, Private Bag, Lower Hutt, New Zealand
General news on DSIR's remote sensing capability, reports on various group activities, conference notices and a brief profile of eminent visitors to DSIR.

346 *E35 Télédétection* M
CDST - CNRS, 26 rue Boyer, F-75971 Paris Cedex 20, France

347 *Earth Observation Quarterly* Q
ESA Publications Division, 'EOQ', ESTEC - PB 299, NL-200 AG Noordwijk, The Netherlands
A useful publication that concerns itself with developments in various aspects of remote sensing and activities within the ESA's Earth Observation Programme.

348 *Earth Resources Mapping in Africa* Q
Regional Centre for Services in Surveying, Mapping and Remote Sensing, Regional Remote Sensing Facility, Box 1818, Nairobi, Kenya
Published since 1979, this bulletin provides general information about remote sensing and the services of the Regional Remote Sensing facility.

349 *EARSeL News* Q
EARSeL Secretariat, 148, rue du fg. St Denis, P.O. Box 60, F-75462 Paris Cedex 10, France

A superb and very comprehensive newsletter. A wide range of topics is covered, including EARSeL association news, members' news, discussion topics and reviews of current satellite technologies. Additionally there are several excellent review sections on worldwide remote sensing organizations, industrial products and services, books and other publications. Symposia, conferences and workshops are reported on, and advance notice is given of forthcoming events. Much of the material that is featured has been abstracted from other serial publications.

350 *EOSAT Landsat Application Notes* bi-M
Earth Observation Satellite Company (EOSAT), 4300 Forbes Boulevard, Lanham, MD 20706, U.S.A.
Prepared for customers to acquaint them with the various commercial applications of Landsat data.

351 *EOSAT Landsat Data Users Notes* Q
Earth Observation Satellite Company (EOSAT), 4300 Forbes Boulevard, Lanham, MD 20706, U.S.A.
Contains news and articles on government views, Landsat receiving station status, purchasing policies and data applications.

352 *EOSAT Landsat Technical Notes* I
Earth Observation Satellite Company (EOSAT), 4300 Forbes Boulevard, Lanham, MD 20706, U.S.A.
This newsletter features technical up-dates on the Landsat programme and information on research and development in image processing and digital image systems.

353 *ESA Bulletin* Q
European Space Agency, 8-10 rue Mario Nikis, F-75738 Paris Cedex 15, France
Provides information on ESA's activities and purposes in an easily assimilated non-technical form. Project reports and announcements of new ESA publications are included and occasional special editions are devoted to particular topics.

354 *ESA Newsletter*
European Space Agency, 8-10 rue Mario Nikis, F-75738 Paris Cedex 15, France
Provides a general summary of events occurring within, or affecting ESA.

355 *European Space Operations Centre Annual Report*
European Space Operations Centre, Robert-Bosch-Strasse 5, D-6100 Darmstadt, Federal Republic of Germany

356 *EUROSENSE Newsletter* Q
EUROSENSE-BELFOTOP N.V., J. Vander Vekenstraat 158, B-1810 Wemmel, Belgium

A glossy publicity mailshot that advertises the services and facilities offered by the commercial remote sensing company EUROSENSE.

357 *Fjarränalys (information frän Rymdbalaget)* bi-M
Svenska Rymdaktubologet, Albygaten 107, S-171 54 Solna, Sweden
News of satellite and sensor status, together with advance conference notices.

358 *Geografia Teorética* bi-M
Instituto de Geociências e Ciências, UNSEP, Rua 10, 2527, Caixa Postal 178, Rio Claro, São Paulo 13500, Brazil
Newsletter of the Associada de Geografia Teorética, published since 1940.

359 *Geosat News* I
The Geosat Committee Inc., 153 Kearny Street, Suite 209, San Francisco, CA 94108, U.S.A.
This informative newsletter contains news on topics that include political developments in remote sensing, Geosat Committee news, projects and issues, articles, political news worldwide and an extensive list of forthcoming events and publications.

360 *IAPR Newsletter* I
International Association for Pattern Recognition (IAPR), Department of Physics Astronomy, University College London, Gower Street, London WC1E 6BT, England
Organizational and individual membership lists and addresses are provided, together with an excellent and somewhat extensive bibliography of reports and literature.

361 *Institut Géographique National: Annual Report*
Institut Géographique National, 2 avenue Pasteur, F-94260 Saint-Mandé, France

362 *Institut Géographique National. Bulletin d'Information* bi-M
Institut Géographique National, Service de la Documentation Géographique, 2 avenue Pasteur, F-94260 Saint-Mandé, France
Contains a well balanced mixture of articles highlighting new technologies and applications in remote sensing and geography. Brief reports on research in progress and a bibliography of publications are included. In French.

363 *Israel Remote Sensing Information Bulletin* I
Interdisciplinary Centre for Technology Analysis and Forecasting (ICTAF), Tel Aviv University, RAMAT AVIV, 69 978 Tel Aviv, Israel
An excellent and very comprehensive newsletter that has a current awareness section on journal contents called the 'ICTAF Bibliography'. Also featured are Israeli-authored papers, international news and advance warnings of conferences, seminars, and courses run at the ICTAF.

364 *La Lettre du CNES* M

CNES, Département de Publications du CNES, 18 avenue Edouard Belin, F-31030 Toulouse Cedex, France

This letter-style publication contains news on the policies and decisions at CNES and industrial and scientific programmes, together with data on developments, techniques, promotional and commercial activities and practical information.

365 *Land and Water* Q

Food and Agriculture Organization of the United Nations (FAO), Remote Sensing Centre (AGLT), Via delle Terme di Caracalla, I-00100 Rome, Italy

This technical newsletter features occasional reports on the progress of the remote sensing section.

366 *LARS News Letter* bi-M

Laboratory for Application of Remote Sensing, Purdue University, Entomology Building - Room 214, West Lafayette, IN 47907, U.S.A.

367 *Maritime Remote Sensing Newsletter* bi-A

Maritime Remote Sensing Committee, P.O. Box 310, Amherst, Nova Scotia B4H 3Z5, Canada

Each edition of this serial is in English and French, and features news, regional reports, histories of organizations, some notes on applications, and a list of forthcoming events.

368 *Meteosat Image Bulletin* M

European Space Operations Centre, Robert-Bosch-Strasse 5, D-6100 Darmstadt, Federal Republic of Germany

Reports on the latest developments in the Meteosat Exploitation Project run by the ESA.

369 *NASA Activities* M

National Aeronautics and Space Administration, Office of Public Affairs, Code LFD-10, FOB 6, Washington, DC 20402, U.S.A.

370 *NASA News* Q

National Aeronautics and Space Administration, Office of Public Affairs, Code LFD-10, FOB 6, Washington, DC 20402, U.S.A.

371 *National Remote Sensing Centre Newsletter* Q

National Remote Sensing Centre, Space Department, Royal Aircraft Establishment, Farnborough, Hampshire GU14 6TO, England

A well produced, yet irregular, publication that provides information on remote sensing developments and activities at the NRSC and the United Kingdom generally.

372 *NCIC Newsletter*

National Cartographic Information Center (NCIC), U.S. Geological Survey, 507 National Center, Reston, VA 22092, U.S.A.

This irregular publication contains news of political developments, meeting notices and a list of NCIC headquarters.

373 *National Space Science Data Center Newsletter* bi-M

National Space Science Data Center (NSSDC), Code 633, NASA/Goddard Space Flight Center, Greenbelt, MD 20771, U.S.A.

Some interesting comments on space policy are featured together with information on the latest developments in computer retrieval at the NSSDC.

374 *News and Views* Q

ESRIN, Via Galileo Galilei, I-00044 Frascati (Rome), Italy

An interesting newsbrief that reports on developments in information retrieval and online databases.

375 *News from Prospace* 3 p.a.

Prospace, 2 place Maurice Quentin, F-75039 Paris Cedex 1, France

A well produced publication presenting the latest news on application programmes, launchers, facilities, products, services and new members. English or French versions are available.

376 *OEEPE UK Newsletter* I

European Organization for Photogrammetric Research (OEEPE), Department of Photogrammetry and Surveying, University College London, Gower Street, London WC1E 6BT, England

A list of the committee members is provided in this newsletter, together with reports on projects in progress, projects being defined, and publications by organization members.

377 *Optical Engineering Reports* I

Society for Photo-optical Instrumentation Engineers, P.O. Box 10, Bellingham, WA 98227, U.S.A.

A useful newsletter that reports on society activities and gives advance notice of forthcoming events.

378 *PEL Remote Sensing Newsletter* I

Remote Sensing Section, Physics and Engineering Laboratory (PEL), Department of Scientific and Industrial Research (DSIR), Private Bag, Lower Hutt, New Zealand

An excellent newsletter that is distributed in New Zealand and the Pacific islands.

379 *Photogrammetric Coyote* Q

E. Coyote Enterprises Inc., Route 4, Building 228, Box 1119, Mineral Wells, TX 76067, U.S.A.

A very well-put-together newspaper-style publication with a lighthearted editorship. News, advertisments, equipment lists and tests, and advance conference notices are included.

380 *Polar Orbiter Newsletter* bi-weekly

Polar Orbiter Newsletter, P.O. Box 8313, Temple Hills, MA 20748, U.S.A.

A very useful service that reports on the imagery that has been acquired by Polar Orbiters, and provides an indication of the data quality for particular application areas.

381 *PST Newsletter* I

ERS-DC, Product Support Team (PST), Project Office, Q134 Building, Space Department, Royal Aircraft Establishment, Farnborough, Hampshire GU14 6TD, England

This new publication provides up-dates of progress made on ERS-1 and lists the PST membership.

382 *Remote Sensing* bi- A

National Environmental Research Council (NERC), Holbrook House, Station Road, Swindon SN1 1DE, England

Contains information and news on NERC remote sensing projects, together with lists of forthcoming conferences and more general news items.

383 *Remote Sensing Association of Australia Newsletter* Q

Remote Sensing Association of Australia, Centre for Remote Sensing, University of New South Wales, P.O. Box 1, Kensington, N.S.W. 2033, Australia

384 *Remote Sensing for Southern Africa* 3 p.a.

Council for Scientific and Industrial Research (CSIR), P.O. Box 395, Pretoria, 0001 Republic of South Africa

Features general and professional news, conference notices and reports on projects at the Satellite Remote Sensing Centre.

385 *Remote Sensing in Canada* Q

Canada Centre for Remote Sensing, Department of Energy, Mines and Resources, 2464 Sheffield Road, Ottawa K1A 0Y7, Canada

This well-presented and informative bulletin presents news on developments in Canadian remote sensing projects, as well as comprehensive reviews and advance notice of symposia and conferences. Each version has the same content printed back-to-back in English and French.

386 *Remote Sensing Nieuwsbrief* Q

Remote Sensing Nieuwsbrief, Postbus 5023, NL-2600 GA Delft, The Netherlands

Published in Dutch, with occasional English articles, this newsletter features news, conference notices and some technical papers on remote sensing. A very useful bibliography of publications and a current awareness service on journal contents is provided by an occasional supplement, the *Bibliotheek bijlage remote sensing nieuwsbrief*, which is produced by the NIWARS-Library, P.O. Box 241, NL- 6700 AE Wageningen, The Netherlands.

387 *Remote Sensing Society's News and Letters* bi-M

Taylor & Francis Limited, Rankine Road, Basingstoke, Hampshire RG24 0BR, England

Reprinted from the *International Journal of Remote Sensing*, this publication features short scientific papers that are intended as forerunners and advance notices of research in progress. In addition to the international news section of the parent journal, this newsletter contains Remote Sensing Society news and is available free to all Society members.

388 *RESTEC Newsletter* bi-M

Remote Sensing Technology Centre of Japan, University-Roppongi Building, 7-15-17, Roppongi, Minato Ku, Tokyo 106, Japan

Published in English and featuring general news, technical developments, data prices and marketing information.

389 *Satellite Remote Sensing Centre Newsletter* Q

Council for Scientific and Industrial Research, P.O. Box 3718, Johannesburg, Transvaal 2000, Republic of South Africa

Features sections concerned with news, meetings, data availability, worldwide activities and employment opportunities.

390 *SENSOR*

Centro Interamericano de Fotointerpretación, Carrera 30 No. 47-A-57, Apartado Aereo 53754, Bogotá D.E., Cundinmarca 2, Colombia

391 *Space Research and Technical Information Bulletin* bi - M

Interdisciplinary Centre for Technology Analysis and Forecasting (ICTAF), Tel Aviv University, RAMAT AVIV, 69 978 Tel Aviv, Israel

An excellent newsletter that is both well produced and comprehensive in its coverage. Features include a very useful section that indexes, abstracts and reviews publications, as well as reporting on additions to online database records. The news section covers domestic and international events, with abstracted journal and newsletter reports and a comprehensive and detailed list of projects and contracts currently being run by ESA, NASA and other major remote sensing organizations. A bibliography of recent publications is provided, as are book reviews, conference notices and a current awareness bulletin of international journal publication contents.

392 *SPOT Ann Report*
SPOT Image, 16 bis avenue Edouard Belin, F-31030 Toulouse Cedex, France

393 *SPOTLIGHT* Q
SPOT Image Corporation, 1897 Preston White Drive, Reston, VA 22091-4362, U.S.A.

394 *SPOT Newsletter* bi-A
SPOT Image, 16 bis avenue Edouard Belin, F-31030 Toulouse Cedex, France
This bilingual publication in English and French advertises recent developments in marketing, research and applications of SPOT data. Well printed examples of the imagery are included, as is a useful mini-directory of SPOT receiving stations and data distributors.

395 *SUPARCO Times* Q
Pakistan Space and Upper Atmosphere Research Commission (SUPARCO), Remote Sensing Applications Centre (RESACENT), P.O. Box 3125, Karachi-29, Pakistan

396 *Swedish Space Corporation Newsletter* Q
Swedish Space Corporation, Albygaten 107, S-171 54 Solna, Sweden

397 *TAC Newsletter* I
Technology Application Center, University of New Mexico, 2500 Central Southeast Albuquerque, NM 87131, U.S.A.
Highlights developments of the computerized facilities at the TAC.

398 *Távérzé Kelési Korlevél* bi-M
Földmérési Intézet (FÖMI), Remote Sensing Centre, P.O. Box 546, Guszev u.19, 1051 Budapest 5, Hungary
A limited-circulation publicity mailshot.

399 *TRSC Newsletter* Q
Thailand Remote Sensing Centre, 196 Phahonyothin Road, Bangkok, 10900 Thailand
Separate editions are printed in English and Thai.

400 *Washington Remote Sensing Letter* 22 p.a.
Washington Remote Sensing Letter, 1057-B National Press Building, Washington, DC 20045, U.S.A.
An excellent publication that acts as a forum for the dissemination of news on politics, policies, legislation, current research and activities by government,

commercial companies and Federal agencies in space technology and remote sensing. Advance notices of symposia are included, together with book reviews and details of other publications.

401 *WMO Bulletin* Q
World Meteorological Organization, 41 avenue Guiseppe Motta, CH-1211 Geneva 20, Switzerland

Translations and Dictionaries

402 Albota, M., Filotti, D., Molea, O. and **Salaria, I.** *Dictionar poliglot di geodezie, fotogrammetrie di cartografie*. Bucharest: Editura Technica (Technical Publishing House), 1976.
Produced in English, Romanian, German, French and Russian. The main entry for each of the 4,700 items that are included is in English but terms are indexed in all the other languages.

403 *Catalogue of NAL Technical Translations* A
National Aeronautical Laboratory, Box 1779, Kodshalli, Bangalore 560017, India

404 Dubios, F.P. *Dictionary of remote sensing terms, English-Spanish, Spanish-English*. Palomar-San Isidro: ONERN, 1980.
A useful translation aid that has been prepared by the Oficina Nacional de Evaluación de Recursos Naturales.

405 Dyson, S. *French-English glossary on SPOT, remote sensing and their application*. 'Bouygues', Miramont-de-Quercy, F-82190 Bourg-de-Visa, France: S. Dyson, 1986. 300pp.
This working document fills a gap in the literature of remote sensing and is a most welcome addition to the dictionaries that are currently available. There are over 3,500 entries, ten separate sections on various remote sensing applications, useful appendices and very helpful cross-referencing to source documents. In keeping with the modern approach of this publication the data is also available on a range of word processing disks.

406 International Society for Photogrammetry. *Multilingual dictionary for photogrammetry*. Amsterdam: Argus, 1969
This dictionary, although dated, is still very useful. It has been produced in seven volumes, one each for English, French, German, Italian, Polish, Spanish and Swedish, each of which cross-references a word in that volume's language to the same word in the other language volumes, thus the series must be used as a set. It has 4,259 word entries.

407 Ismailor, T. K. *English-Russian dictionary on remote sensing of the natural earth resources.* Academy of Sciences of the Azerbaijan S.S.R.: The Scientific Centre 'Caspij' Publishing House 'Elm', 1977. 363pp.
Four thousand entries.

408 Konarski, M. M. *Russian-English space technology dictionary.* London: Pergamon Press, 1980. ISBN 08-015617-7
Over 10,600 selected terms used in space technology are included, together with an index of English and referred terms.

409 *National Aeronautics and Space Administration, Technical Translations* I
National Technical Information Service, 5285 Port Royal Road, Springfield, VA 22151, U.S.A.

410 Paul, S. *Dictionnaire de télédétection aérospatiale.* Paris: Masson, 1982. 236pp. ISBN 2-2257-5889-1
The terms and definitions in this dictionary are in French, although introductory material and indexes are in both French and English.

411 Rabchevsky, G., ed. *Multilingual dictionary of remote sensing and photogrammetry.* Falls Church, Virginia: American Society of Photogrammetry, 1984. 386pp.
This work contains 1,716 terms defined in English and numbered and translated into French, German, Italian, Spanish, Portuguese and Russian. Terms in each language index are alphabetically listed and cross-referenced to the English definitions, and there is a bibliography and list of acronyms.

412 Taillefer, Y. *Recueil de terminologie spatiale ESA SP-1011.* Paris: European Space Agency, 1982.

413 *U.S.S.R. Report: EARTH SCIENCES. Abstracts only* I
Joint Publications Research Service, Arlington, VA 22161, U.S.A.
An occasional translation service which covers the whole range of earth sciences, including remote sensing and meteorology.

414 *World Transindex* 10 p.a.
International Translations Center, Doelenstraat 101, NL-2611 NS Delft, The Netherlands
This journal announces translations received in all fields of science and technology, and is prepared in English, French and German.

Regular Conferences

415 *Annual Convention*
416 *Fall Convention*

American Congress on Surveying and Mapping (ACSM)/American Society of Photogrammetry and Remote Sensing (ASPRS), 210 Little Falls Street, Falls Church, VA 22046, U.S.A.

The Annual Convention is held in March and its proceedings are generally named *Technical papers from the Annual Meeting*. The Fall Convention is normally in September or October and its proceedings are similarly named *Technical papers from the Fall Convention*. The ASPRS also sponsors workshops and seminars on various themes in remote sensing on an irregular basis.

417 *Biennial Workshop on Color Aerial Photography in the Plant Sciences and Related Fields*

American Society of Photogrammetry and Remote Sensing (ASPRS), 210 Little Falls Street, Falls Church, VA 22046, U.S.A.

Run biennially since 1967, this workshop deals with the acquisition, processing and interpretation of aerial photographs. The basic methodologies and applications of remote sensing are clearly documented.

418 *Annual Convention*

419 *Spring Convention*

American Society of Civil Engineers (ASCE), 345 East 47th Street, New York, NY 10017, U.S.A.

Although not devoted to remote sensing, some of the sessions sponsored by the Engineering Surveying Division feature papers on photogrammetry and remote sensing, generally in an engineering application.

420 *Canadian Symposium on Remote Sensing*

Canada Centre for Remote Sensing, Department of Energy, Mines and Resources, 2464 Sheffield Road, Ottawa, Ontario K1A 0Y7, Canada

An annual symposium that reports on the developments in Canadian remote sensing.

421 *Congreso Nacional de Fotogrametría, Fotointerpretación y Geodesia*

Mexican Society of Photogrammetry, Photointerpretation and Geodesy, Camino A. Sta Teresa 187, Mexico 22, D.F.

The proceedings of this biennial congress are published as *Memorias, Congreso Nacional de Fotogrametría, Fotointerpretación y Geodesia*.

422 *International Symposium on Acoustic Remote Sensing of the Atmosphere and Oceans*

Consiglio Nazionale delle Richerche, Piazzale delle Scienze 7, I-00185 Rome, Italy

423 *EARSeL/ESA Symposium*

EARSeL Secretariat, 148 rue du Fg. Saint Denis, P.O. Box 60, F-75462 Paris Cedex 10, France

424 *International Symposium on Remote Sensing of the Environment*
Environmental Research Institute of Michigan, Ann Arbor, MI 48107, U.S.A.
A very varied programme of papers is presented at this annual conference. Proceedings contain keywords, author affiliation lists and excellent indexes. Several concurrent sessions on specialized topics are run during the conference.

425 *Thematic Conference on Remote Sensing for Exploration Geology at the International Symposium on Remote Sensing of the Environment*
Environmental Research Institute of Michigan, Ann Arbor, MI 48107, U.S.A.
One of the concurrent sessions from the International Symposium on Remote Sensing of the Environment that has emerged as a regular conference in its own right.

426 *Geosat Workshop: Frontiers of Geological Remote Sensing from Space*
Geosat Committee Inc., 153 Kearny Street, Suite 209, San Francisco, CA 94108, U.S.A.
This meeting presents state-of-the-art research and development papers on the geological applications of remotely sensed data. These cover a wide spectrum of geologic interest and contributions are presented from each of the six Geosat working groups.

427 *ICAP - International Conference on Antennas and Propagation*
Institution of Electrical Engineers (IEE), Savoy Place, London WC2R 0BL, England

428 *ICASSP - IEEE International Conference on Acoustics, Speech and Signal Processing*
Institute of Electrical and Electronics Engineers Inc. (IEEE), 345 E. 47th Street, New York, NY 10017, U.S.A.

429 *IEEE Computer Society Conference on Computer Vision and Pattern Recognition*
Institute of Electrical and Electronics Engineers Inc. (IEEE), 345 E. 47th Street, New York, NY 10017, U.S.A.

430 *IEEE International Conference on Pattern Recognition*
Institute of Electrical and Electronics Engineers Inc. (IEEE), 345 E. 47th Street, New York, NY 10017, U.S.A.

431 *IEEE International Microwave Symposium and Workshops*
Institute of Electrical and Electronics Engineers Inc. (IEEE), 345 E. 47th Street, New York, NY 10017, U.S.A.

432 *IGARSS. IEEE Geoscience and Remote Sensing Society Symposium*

Institute of Electrical and Electronics Engineers Inc. (IEEE), 345 E. 47th Street, New York, NY 10017, U.S.A.

433 *ISPRS Quadrennial Congress*
434 *ISPRS Technical Commission 1 Symposium*
ISPRS Technical Commission 2 Symposium
ISPRS Technical Commission 3 Symposium
ISPRS Technical Commission 4 Symposium
ISPRS Technical Commission 5 Symposium
ISPRS Technical Commission 6 Symposium
ISPRS Technical Commission 7 Symposium
International Society for Photogrammetry and Remote Sensing, Department of Photogrammetry, S-100 44 Stockholm, KTH, Sweden
The ISPRS Quadrennial Congress proceedings are published as the *International Archives of Photogrammetry and Remote Sensing*. Additionally each of the Technical Commissions holds a conference during the four-year period between the major Quadrennial Congresses, and the proceedings of each of these is published by the sponsoring member as volumes of the *Archives*.

435 *Auto-Carto*
International Cartographic Association, Flottbrov 16, S-112 64 Stockholm, Sweden
This conference concentrates on reviewing the latest spatial information extraction, collection, analysis and processing methodologies and applications.

436 *International Symposium on Machine Processing of Remotely Sensed Data*
Laboratory for Applications of Remote Sensing, Purdue University, Entomology Building - Room 214, West Lafayette, IN 47907, U.S.A.
The annual conferences usually deal with a specific theme in remote sensing. Most of the contributors originate from within the U.S.A. An excellent series that reports on the most up-to-date research in remote sensing.

437 *Annual Conference of the Remote Sensing Society*
Remote Sensing Society, Department of Geography, University of Nottingham, University Park, Nottingham NG7 2RD, England
The major conference held in the United Kingdom on remote sensing. Each meeting concentrates on a different aspect of remote sensing and has contributions of a markedly international flavour.

438 *Australasian Remote Sensing Conference*
Remote Sensing Association of Australia, Centre for Remote Sensing, University of New South Wales, P.O. Box 1, Kensington, N.S.W. 2033, Australia
Every three years, the most recent being in 1987.

439 *Annual Information Seminar*

Regional Centre for Services in Surveying, Mapping and Remote Sensing, P.O. Box 18118, Nairobi, Kenya

440 *Annual International Symposium*
Society for Photo-optical Instrumentation Engineers (SPIE), P.O. Box 10, Bellingham, WA 98227, U.S.A.
Held in August. Concurrent sessions are given on various specialized topics such as the application of digital image processing to remote sensing and reviews of remote sensing technology. The Society also sponsors around fifty other conferences, symposia and workshops every yea.; these are available as *Proceedings of the SPIE.*

441 *Annual SPSE Conference*
Society of Photographic Scientists and Engineers (SPSE), 7003 Kilworth Lane, Springfield, VA 22151, U.S.A.
The Annual Conference is held in May and the Fall Symposium in October or November. Specialized workshops on electronic and optical imaging technology are held periodically.

442 *Pecora Symposium*
U.S. Geological Survey, 917 National Center, Reston, VA 22092, U.S.A.
The Pecora Symposia deal with the most recent reports on space remote sensing developments. Contributions are from leading individuals in commercial organizations, industry, government and Federal agencies. The political and commercial implications of remote sensing are given a good airing.

Source Materials for Remotely Sensed Imagery

443 **Dodd, K., Fuller, H. K.** and **Clarke, P. F.** *Guide to Obtaining USGS Information: USGS Circular 900*
Books and Open-File Reports Section, U.S. Geological Survey, Box 25425, Federal Center, Denver, CO 80225, U.S.A.
A major revision of former USGS information guides that includes a useful description of the recently reorganized USGS distribution and user services. USGS products are exhaustively tabulated, and their availability (whether for loan, hire, inspection, or mail) is clearly indicated. Amongst a wealth of other items the guide reports on aerial photographs, satellite images, maps, motion pictures, digital data, indexes and catalogues. Descriptions of several USGS centres are included, possibly the most useful being those of the National Cartographic Information Center and USGS photographic library.

444 **ESA.** *Earthnet - the story of images: BR-18.* Paris: European Space Agency, 1984. 59pp.
A dated but useful guide to the activities of the former Earthnet network.

445 *General Coverage Systems Catalogue (Catalogue de la Couverture Photographique Aérienne du Canada)*
Surveys and Mapping Branch, National Air Photo Library, 615 Booth Street, Ottawa, Ontario K1A 0E9, Canada

This catalogue provides a comprehensive description of the photographic products available from the National Air Photo Library of Canada. The introduction describes the National Topographic System and its use, while the rest of the publication details the availability of catalogues of different areas of Canada, microfilmed index maps and imagery, ordering information and addresses, products (industry contact prints, transparencies, enlargements, diapositives and mosaics) and miscellaneous services.

446 High Schools Geography Project. *High Schools Geography Project: sources of information and materials; maps and aerial photographs.* Association of American Geographers, Committee on Maps and Aerial Photographs, 1970.

This guide to resources, although somewhat dated and biased toward American information sources, provides a useful list of maps, aerial photographs and satellite imagery, together with details of slide sets and filmstrips, and the addresses from which all these products may be obtained.

447 Holt, B. *Seasat SAR - Imagery Catalog.* Pasadena: Jet Propulsion Laboratory, 1982.

A complete record of all imagery acquired by the Seasat SAR satellite. Areal maps of all Seaset swaths that satisfy certain data quality requirements are included, together with extensive numerical tables that list the various parameters of data collection. Section IV is particularly interesting as it lists all of the images that were processed by the Jet Propulsion Laboratory up until October 1981, providing notes on their latitude and longitude, together with a description of geographical location.

448 Kroeck, D. *Everyone's space handbook: a photo imagery source manual.* Arcata California: Pilot Rock, 1976. 175 pp.

A brief introductory guide to remote sensing is included in this manual, which lists the agencies and organizations that archive and disseminate image data.

449 *LANDSAT MicroCatalog: a reference system.* 1983
NOAA/NESDIS Landsat Customer Services, Mundt Federal Building, Sioux Falls, SD 57198, U.S.A.

This user-aid contains a thorough explanation of the Worldwide Reference System for Landsat Data, together with brief explanatory notes on the use of the catalogue. The concise text is complemented by very useful illustrations that clearly explain the structure and function of the Landsat system, and the book itself.

450 NASA. *Catalog of Space Shuttle Earth Observation Hand-Held Photography. Space Transportation Systems (STS) 1, 2, 3 and 4 Missions. JSC-20039*. Houston: NASA Johnson Space Center, 1985.

451 NASA. *Heat Capacity Mapping Mission (HCMM) data users' guide for applications - Explorer mission A*. Greenbelt, Maryland: Goddard Space Flight Center, 1980. 120pp.
The third up-date in the NASA HCMM application series.

452 NASA: Lyndon B. Johnson Space Center. *Skylab EREP Investigations Summary*. Washington D.C.: NASA, 1978. 386pp.
A summary of the Earth Resources Experiment Package (EREP) that tested the use of visible, infrared and microwave sensors in monitoring and studying terrestrial resources. Research areas include land use and cartography, agriculture, range and forestry, geology and hydrology, oceans, atmosphere and data analysis techniques. References are provided for each topic and the back of this summary contains a list of sensor systems, principal investigators and a coverage index. A particularly useful section deals with the principles of photographic and digital data analysis.

453 NASA. *Skylab Earth Resources Data Catalog*. Houston: Johnson Space Center, 1974. 393pp. NASA-TM-X-70411
Original colour imagery is featured in this manual, which presents an overview and index of Skylab photography. The question of using Skylab data for various applications is discussed and encouraged. This book also provides a complete index of the Skylab Earth Resources Experiment Package (EREP) photographs. Examples are given, together with explanations of the potential uses of the more than 35,000 frames of imagery and 73,415m of taped data taken on this mission.

454 NASA. *The Nimbus-7 users guide*. Greenbelt, Maryland: Goddard Space Flight Center, 1978. 263pp.
A unique and interesting guide to the imagery and applications from this satellite series.

455 *NRSC Data Users' Guide* semi-A up-dates
National Remote Sensing Centre, R190 Building, Space Department, Royal Aircraft Establishment, Farnborough, Hampshire GU14 6TD, England
This very well presented guide not only describes the activities of the NRSC and reviews major spaceborne remote sensing systems, but also provides a set of index maps for Landsats 1, 2, 3, 4 and 5, and SPOT that cover the whole world. Informative appendices list the data holdings of the NRSC and also provide information on prices and products. Although this publication is designed for UK users, it can be strongly recommended as a very comprehensive general guide to worldwide remote sensing systems and their coverage.

456 *Reports of an expert consultation on a World Index of Space Imagery (WISI), World Aerial Photography Index (WAPI) and thematic cartography for renewable natural resources development* I
Food and Agriculture Organization, Remote Sensing Centre (AGLT), Via delle Terme di Caracalla, I-00100 Rome, Italy

A very useful report series that reviews the progress made on this worthwhile enterprise. A number of position papers are usually presented by individual countries, and there are often contact addresses and flight-maps that give an indication of the photographic resources of developing nations.

457 Scott Southworth, C. *Characteristics and availability of data from earth-imaging satellites. U.S. Geological Survey Bulletin 1631.* 1985. 102pp.

A very useful guide to the coverage obtained by a range of spaceborne systems, including Landsat, Heat Capacity Mapping Mission, Seasat and SIR-A. The index maps and images (some of which are from the Landsat TM) that are included tend to focus on North America. The brief accompanying narrative is kept to a minimum, and there are two appendices, which respectively list optically processed and digitally processed Seasat data. A short and concise bibliography is included, together with a selected list of satellite data information suppliers.

458 Short, N.M. and **Stuart, L.M.** *The Heat Capacity Mapping Mission (HCMM) anthology. NASA SP-465.* 1983. 264pp. U.S. National Aeronautics and Space Administration, Washington, DC 20546, U.S.A.

This very large-format publication presents a superb review of the HCMM that runs from the historical perspective of the programme to a summary and overview. The results from the programme are described, together with interpretative comments on some of the imagery. The illustrations are excellent, being backed up by line drawings of maps and neatly summarized information tables.

Audio-Visual Materials

Training and Reference Manuals for Remotely Sensed Data

459 NASA. *Landsat Data Users Handbook.* 1984.
Distribution Branch, U.S. Geological Survey, 604 South Pickett Street, Alexandria, VA 22304, U.S.A.

A very useful technical guide to the Landsat system that may be kept current by reference to *EOSAT Landsat Data Users Notes, EOSAT Landsat Application Notes*, and *EOSAT Landsat Technical Notes.*

460 Sabins, F. *Instructor's key for remote sensing laboratory manual.* La Habra, California: Remote Sensing Enterprises, 1981. *c.* 100pp; in various paginations.

A useful answer-book and source of pertinent information for the main guide book.

461 Sabins, F. *Remote sensing laboratory manual.* 2nd ed. La Habra, California: Remote Sensing Enterprises, 1981. *c.* 300pp; in various paginations.
An excellent loose-leaf manual that complements an accompanying set of slides. Pagination is divided into sections, which feature maps and overlays, diagrams and exercises, together with information on Landsat-4 TM and the SIR-A experiment.

Maps, Map Catalogues and Guide Books

462 *Africa: Guide to available mapping* I
McCarta Ltd., 122 Kings Cross Road, London WC1X 9DS, England
Apart from a very comprehensive coverage of topographic maps for various areas of Africa, this guide lists well over 100 photomaps.

463 *Americas: Guide to available mapping* I
McCarta Ltd., 122 Kings Cross Road, London WC1X 9DS, England
Several U.S. Geological Survey satellite image maps are noted on this occasional list, as well as nine photogeological maps and nearly seventy photo-maps.

464 *Asia, Australia, Pacific and Indian Oceans: Guide to available mapping* I
McCarta Ltd., 122 Kings Cross Road, London WC1X 9DS, England
Several satellite maps, coloured photo-maps and other photo-maps are identified by this publication.

465 *Europe: Guide to available mapping* I
McCarta Ltd., 122 Kings Cross Road, London WC1X 9DS, England
Although this list only identifies one satellite image map, it covers other sorts of maps very thoroughly and it is worth consulting both this and future editions.

466 Hodgkiss, A. G. and **Tatham, A. F.** *Keyguide to information sources in cartography.* London: Mansell Publishing, 1986. 253pp. ISBN 0-7201-1768-2

467 *International Yearbook of Cartography* A
London: George Philip & Son.
Dominantly concerned with map-making and cartography, but still of interest to those involved in remote sensing.

468 *Modern maps and atlases: an outline guide to twentieth century production.*
London: Clive Bingley, 1969. 619pp.
A dated yet comprehensive guide to thematic and topographic maps and mapping.

469 *BANGLADESH - landcover indications derived from Landsat imagery*, 1981.
Map III (Base Map - two colour: light brown and black); utilizing portions of 14 scenes; scale 1:500,000.

470 *BANGLADESH - landcover, soil and water reflections from Landsat imagery*, 1981. Map I; utilizing portions of 14 scenes; scale 1:500,000.

471 *BANGLADESH - land use classification related to landcover reflections obtained from Landsat imagery*, 1981. Map II; utilizing portions of 14 scenes; scale 1:500,000.

472 *BHUTAN - Landsat, soil, and water reflections from Landsat imagery*, 1982. (Edge-enhanced false-colour composite); utilizing portions of 5 scenes; scale 1:250,000.

473 *BURMA - land cover-land use association*, 1976. (Based on computer tape analysis of Landsat data); map shows 80% of the country's coastline and more than 35% of the country's area; utilizing portions of 14 scenes; scale 1:1,000,000.

474 *HAUTE-VOLTA (UPPER VOLTA) - zones à potentiel de développement (potential settlement areas)*, 1977. Project de Développement Agricole de l'Ouest Volta (West Volta Agricultural Development Project); portions of 2 scenes; scale 1:250,000.

475 *NEPAL - landcover indications derived from Landsat imagery*, 1982. (Edge-enhanced false-colour composite); utilizing portions of 14 scenes; scale 1:500,000; set has two map sheets: I Western Sheet, II Eastern Sheet.

476 *NEPAL - landcover indications derived from Landsat imagery*, 1982. (Base map - two colour: light brown and black); utilizing portions of 14 scenes; scale 1:500,000; set has two map sheets: I Western Sheet, II Eastern Sheet.

477 *ORISSA, INDIA - land cover-land use association*, 1977. (Based on computer tape analysis of Landsat data); Wet Season utilizing portions of 13 scenes; scale 1:1,000,000.

478 *PERU - Selva Central (Zona Sur y Zona Norte)*, 1982. Mosaico de Imagenes del Satelite Landsat; utilizing portions of 4 scenes; scale 1:100,000; set has two map sheets: South Zone and North Zone.

Atlases and Manuals of Remotely Sensed Imagery

479 *Atlas of false colour Landsat images of China*. Beijing: Science Press (distributed by Asian Research Service: Hong Kong), 1984. 553pp. (3 volumes).

A very large and unique atlas that presents black-and-white and false-colour composite Landsat 1, 2 and 3 images of China's territory and border regions. The images are at a scale of 1:500,000, and the atlas's text is in both English and Chinese.

480 Bodechtel, J., ed. *Weltraumbild Atlas, Deutschland, Österreich, Schweiz.* Braunschweig: Georg Westermann Verlag, 1978. 88pp.

The imagery in this publication is at 1:500,000 scale, which helps in its comparison with the accompanying maps.

481 Bullard, R. K. and **Dixon-Gough, R. W.** *Britain from space: an atlas of Landsat images.* London and Philadelphia: Taylor & Francis, 1985. 128pp. ISBN 0-85066-277-X

The glossy format of this atlas belies the highly generalized nature of its content, and considering the title, this publication is let down by the imagery displayed. It is, however, useful for schools and colleges as it contains a directory of academic institutions, government bodies, industrial and commercial organizations. A glossary and list of abbreviations are also featured.

482 Dickison, P. *Out of this world.* New York: Delaconte Press, American Space Photography, 1977. 158 pp. ISBN 0-440-06568-2.

This atlas, which includes fifty-four panchromatic and colour plates, presents the early history of space photography and considers Landsat, meteorological and terrestrial images.

483 Ford, J. P., ed. *Seasat views North America, the Caribbean, and Western Europe with imaging radar. JPL Publication 80-67.* Pasadena: Jet Propulsion Laboratory, 1980. 141pp.

484 Ford, J. P., Cimino, J. B. and **Elachi, C.** *Space Shuttle Columbia views the world with imaging radar - the SIR-A experiment. JPL Publication 82-95.* Pasadena: Jet Propulsion Laboratory, 1983. 176pp.

485 Ford, J. P., Cimino, J. B., Holt, B. and **Nuzek, M. N.** *Shuttle imaging radar views the Earth from Challenger - the SIR-B Experiment. JPL Publication 86-10.* Pasadena: Jet Propulsion Laboratory, 1986.

486 Francis, P. and **Jones, P.** *Images of Earth.* Englewood Cliffs, New Jersey: Prentice-Hall, 1984. 159pp. ISBN 0-5400-1083-9

An attractive publication that presents images from past as well as present satellite and spaceborne systems.

487 *Images of the world: an atlas of satellite imagery and maps.* (With interpretive supplement by **Smith, N.M.**). Glasgow: Collins - Longman Atlases, 1984. 175pp. ISBN 0-00-360205-2

A well presented atlas that makes clever use of maps in its introduction to satellite imagery. A brief introduction to remote sensing is made at the beginning and world maps indicate the location of each image, and its potential application area. One of the most striking things about this atlas is the lack of wasted space; colourful images and maps cover every page, and interpretive map symbols are placed in the small white border areas that remain.

488 NASA. *Earth photographs from Gemini III, IV and V. NASA SP-129.* 1967. 266pp. LOC No. 66062089.

Nearly 250 photographs of East Africa, Mexico, the Middle East and the United States are featured in this image catalogue. Many of the photographs were obtained for applications experiments and tend to concentrate on areas of particular geologic, geographic and oceanographic interest. Some technical details are included, and a few of the pictures are briefly annotated.

489 NASA. *Skylab Explores the Earth. NASA SP-380.* 1977. 517pp. LOC No. 77000829

A dominantly application-oriented volume, with each chapter addressing a particular practical problem. The illustrations are uniformly superb and are backed up by appendices that include a glossary, list of weather markings and index of photographs.

490 Sheffield, C. *Earthwatch: a survey of the world from space.* London: Sidgwick & Jackson, 1981. 160pp. ISBN 0-283-98737-5

An excellently packaged catalogue of some of the Earth's major surface features as registered by the Landsat MSS. This general introduction is not meant for technical purposes, but the quality of the images and lucid commentary make it a very informative publication.

491 Sheffield, C. *Man on Earth: the marks of man, a survey from space.* London: Sidgwick & Jackson, 1983. 160pp. ISBN 0-283-98956-4

A companion book to the original *Earthwatch*. The format is the same, but the focus is different, concentrating on the influence man has had on his environment.

492 Short, N. M. *Mission to Earth: Landsat views the world. NASA SP-360.* 1976. 459 pp. LOC No. 76608116

One of the earliest imagery atlases, and the first to present results from the early Landsat missions.

493 World Bank. *Landsat Index Atlas of the Developing Countries of the World.* Baltimore: Johns Hopkins University Press, 1976. 17pp.

A large-format ring-bound publication that has been designed to provide an index of the available satellite coverage for developing countries. The layout, which includes a narrative description of possible applications and image interpretation, is geared towards promoting the use of satellite imagery for less-developed nations.

PART III

Organizational Sources of Information

Organizational Sources of Information

The list of organizations below represent the key groups and institutions involved in remote sensing for the countries specified. The fuller entries are based on questionnaire returns, and the organizations from those addresses that are not annotated are urged to contact the author with information on their activities and services.

Some countries, notably North American and Western European ones, are very well represented, although it has been necessary owing to limited space to exclude some entries. Universities are mostly left out of this directory, as remote sensing groups are very well represented in many, and it would be unrealistic to include all of those here.

The numbers 1 and 2 in the entries below respectively refer to the head of an organization and its chief information officer, or librarian. It should be noted that it is often more prudent to write to the title, rather than the actual name of an individual, as staff changes are fairly frequent.

International

494 European Association of Remote Sensing Laboratories (EARSeL), EARSeL Secretariat, 148 rue du Fg. Saint Denis, P.O. Box 60, F-75462 Paris Cedex 10, France

Tel. 42 05 28 90

1. Dr. L. F. Curtis, Secretary General
2. M. Godefroy, Secretary

Finance: Council of Europe (Division of Education and Research), Commission of the European Communities and the European Space Agency

Purpose: The encouragement and co-ordination of remote sensing activities in Western Europe

Main Interests: All aspects of remote sensing, education and training in remote sensing

Education and Training: EARSeL has fifteen working groups and a membership of 180 research laboratories and sixty observer members from commercial firms and other organizations. The working groups regularly organize workshops and seminars on different topics in remote sensing. Teaching material is being developed for different-level university courses in remote sensing. Annual conference (often held jointly with the ESA)

Publications: EARSeL Directory of members and observers (annual update), *EARSeL News* (quarterly)

495 European Space Agency (ESA), 8-10 rue Mario-Nikis, F-75738 Paris Cedex 15, France
Tel. (1) 273 7654 Telex ESA 202746
1.(a) Dr. R. Lust, Director
(b) P. Goldsmith, Director of Earth Observation and Microgravity Programmes
(c) E. S. Mallet, Director of Applications Programme
(d) Dr. F. Engstrom, Director of Columbus Programme
(e) Dr. R. Bonnet, Director of Scientific Programme
(f) M. Bignier, Director of Space Transportation Systems
(g) G. Salvatori, Director of Telecommunication Programme
Finance: European governments
Purpose: The inter-government space agency for eleven member states. Co-ordination of space research and development and the manufacture of satellites, space platforms, launchers and payloads for Europe
Main Interests: The main remote sensing interests of ESA are concerned with monitoring its own application satellites, Meteosats 1, 2 and P2 (weather satellites), and the maintenance of Ariane (European satellite launch vehicle). As well as research and development of ERS-1 (Earth Observation Satellite-1), the ESA is also involved in several other satellite programmes, namely Exosat, Giotto (Halley's comet investigator), Hipparcos (astronomical satellite), ISO (the Infrared Space Observatory) and, in collaboration with NASA, ISPM (sun pole investigator)

496 European Space Operations Centre, Meteosat Exploitation Project, Robert-Bosch-Strasse 5, D-1600 Darmstadt, Federal Republic of Germany
Tel. 49 6151 886 1 Telex 419453
1. J. de Waard, Project Manager
2. Dr. O. M. Turpeinen, Theoretical Meteorologist
Finance: Inter-government
Purpose: Exploitation of the European Meteorological Satellite (METEOSAT). Also maintenance of all satellite operations, ground facilities and communications, allied with satellite attitude monitoring and control, future project planning
Main Interests: Acquisition, processing and analysis of satellite data and its dissemination through an established network to the user community. Ground facility research, development and maintenance. Meteorology
Education and Training: The different departments host Third World students who take part in various development projects. Meteosat Users Meeting every eighteen months
Library: 14,000 volumes, 245 current periodicals. Complete archive of Meteosat data, digital and photographic, from 1977 onward
Publications: Annual report, also *Meteosat Image Bulletin* (monthly). Staff publish articles and scientific papers

497 European Space Research and Technology Centre, European Space Agency Publications Division, Postbus 299, Noordwijk NL-2200 AG, The Netherlands
Tel. (1719) 83400 Telex 39098
1. M. Le Fevre, Director
2. (a) R. Collete, Director of Earth Observation Programme
(b) N. Longdon, Head of ESA Publications Division
Finance: Inter-government
Purpose: Research, development and testing of spaceborne satellites, systems and payloads
Main Interests: Development of satellite systems for remote sensing
Education and Training: Co-sponsor of the annual EARSeL/ESA Symposium. Other remote sensing workshops, normally one or two a year
Publications: Responsible for all ESA publications, which include the annual report, Proceedings of the annual EARSeL/ESA Symposium, *ESA Journal* (quarterly), *ESA Bulletin* (quarterly), *Earth Observation Quarterly, ESA Special Publications* (irregular), Brochure (irregular), *ESA Scientific and Technical Reports and Memoranda* (irregular), *ESA Procedures, Standards and Specifications* (irregular), *ESA Contractor Reports* (irregular), *ESA Tribology Series* (irregular). Catalogues of ESA publications are available upon request

498 ESRIN, Via Galileo Galilei, I-00044 Frascati (Rome), Italy
Tel. (39/6) 94011 Telex 610 637 ESRIN I
1. F. Roscian, Head
2. Dr. Livio Marelli, Head of Earthnet Programme Office
Finance: Inter-government
Purpose: ESRIN houses two services: the Earthnet Programme Office (EPO), which is responsible for EARTHNET activities, including satellite data acquisition, processing and dissemination, and the Information Retrieval Service (IRS), with the main function of providing an online information facility for the European Space Agency and other establishments requiring this service
Main Interests: Acquisition, archiving, processing and distribution of remote sensing data from different missions. Provision of online services
Library: Maintains the ESA - Information Retrieval Service (IRS), a host that carries over 35 million bibliographic references on over eighty machine-readable databases. Access to 300,000 Landsat, 1,000 Seasat, 3,000 Nimbus, 600 Heat Capacity Mapping Mission and 1,000 Metric Camera images
Publications: News and Views (quarterly)

499 Food and Agriculture Organization, Remote Sensing Centre (AGLT), Agriculture Department, Via delle Terme di Caracalla, I-00100 Rome, Italy
Tel: 00100 57971 Telex 610181 FAO I
1. J.A. Howard, Chief of the Remote Sensing Centre
Finance: United Nations, inter-governmental organizations and contributions from member states
Purpose: To provide a world centre for information and co-operation in agriculture;

the mobilization of capital for agricultural development and contributing to the food security of developing countries
Main Interests: Applications of remote sensing in agriculture, forestry and the fisheries
Education and Training: Ten short-term (2-4 week) training courses on the application of remote sensing to development and management of renewable natural resources in developing countries are run annually
Library: 50 volumes, 10 current periodicals, approximately 1,000 reports proceedings and brochures, slide sets and videos. Selected images from the Space Shuttle, SAR, SLAR and aerial photographs of developing countries
Publications: A list of publications is available on request

500 International Institute for Aerospace Survey and Earth Sciences ITC, 350 Boulevard 1945, P.O. Box 6, NL-7500 AA Enschede, The Netherlands
Tel. 53 320330 Telex 44525
1. Prof. Dr. K. J. Beek, Rector
2. Ing. J. H. Ten Haken, Librarian
Finance: Government/consultancy
Purpose: Training and educating professionals, mainly from developing countries, in surveying and mapping, the acquisition and use of aerial photographs, satellite images and remotely sensed data
Main Interests: Application of small- and large-format aerial photographs for resource surveys, application of SPOT and Landsat MSS/TM data
Education and Training: A wide variety of training courses for postgraduates, experts and technicians; the duration is flexible, and largely dependent on ability and requirement. Full details from the *ITC General Information Booklet*, available on request. Permanent committee on student interests
Library: 15,000 volumes, 500 journals, small AV-production unit for educational activities and over 160 films, slide sets, video and audio cassettes. Forty-one international photo maps are held in the cartographic department of the map library, together with several hundred satellite images and several thousand aerial photographs (worldwide)
Publications: Annual report, *ITC Journal* (quarterly), *ITC General Information Booklet* (annual), *ITC Publications* (irregular); annual staff publication lists are available on request

501 International Society for Photogrammetry and Remote Sensing (ISPRS), Department of Photogrammetry, S-100 44 Stockholm, KTH, Sweden
Tel. 46 8 787 7344 Telex S 10389
1. Prof. Dr. G. Konecny, President of ISPRS
2. Prof. Dr. A. K. I. Torlegård, Secretary General of ISPRS
Finance: Membership fees
Purpose: The development of international co-operation in the advancement of photogrammetry and remote sensing and their applications
Main Interests: All

Education and Training: The ISPRS has seven Technical Commissions which are responsible for education and training, concerned respectively with: primary data acquisition; instruments for data reduction and analysis; acquisition and use of space photographic data; mathematical analysis of data; cartographic and databank applications of photogrammetry and remote sensing; other non-cartographic applications of photogrammetry and remote sensing; economic, professional and educational aspects of photogrammetry and remote sensing; interpretation of photographic and remote sensing data. Each commission holds a Quadrennial Technical Symposium in the years between the ISPRS's major Quadrennial Congresses, the next of which takes place in Kyoto, Japan in 1988.
Publications: Photogrammetria (bi-monthly), *International Archives of Photogrammetry and Remote Sensing* (irregular and dependent on conference frequency)

502 Joint Research Centre (JRC), Commission of the European Communities, I-21020 Ispra (Varese), Italy
Tel. (39 332) 789874 Telex 380042/380058 EURI
1. G. Fraysse, Remote Sensing Programme Manager
Finance: International
Purpose: Research centre of the European Communities
Main Interests: Application of remote sensing techniques to marine pollution, land use and agriculture in European Economic Community (EEC) and Sahelian countries
Education and Training: A variety of specialist courses are run in land use, synthetic aperture radar, crop production forecasting, image processing, agriculture and hydrology. Course duration is from one to three weeks and some are run in collaboration with the ESA, EARSeL and the Council of Europe. Leaflets detailing the course content are available from the Joint Research Centre. Specialized workshops and seminars take place on an irregular basis and cover such fields as thermal inertia, atmospheric corrections, hydrodynamic marine modelling, oil detection, oil fluorescence and coastal pollution
Library: 58,000 volumes, 1,200 current periodicals. Many hundreds of Landsat MSS and TM, NIMBUS, NOAA Thermal MSS, SPOT simulation and SAR-580 images of the EEC and Sahel
Publications: Bi-annual report, final report for each project, reports on technical activities, technical notes and communications. A list of technical papers and articles produced by the staff for journal and conference publication is available on request

503 World Meteorological Organization (WMO), 41 avenue Guiseppe Motta, CH-1211 Geneva 20, Switzerland
Tel. (022) 34 64 00 Telex 23260 OMM
1. M. Favre, Librarian
2. R. Mathieu, Public Information Officer
Finance: United Nations budget

Purpose: To facilitate worldwide co-operation in the establishment of networks for making meteorological as well as hydrological and other geophysical observations, and to establish centres for the provision of meteorological services and observations. The encouragement of research and training in meteorology and other related fields

Main Interests: Application of remote sensing to meteorology and operational hydrology

Education and Training: Specialized courses in co-operation with WMO members and other international organizations. Remote sensing training data are part of an integrated set of such material held by the WMO Secretariat

Library: 26,000 volumes, 240 current periodicals

Publications: Annual report, also *WMO Bulletin* (quarterly) in English, French, Russian and Spanish. A catalogue of publications is available, as are annual supplements which supersede the previous year's up-date

504 Asian Association on Remote Sensing Secretariat (AARS), Institute of Industrial Science, University of Tokyo, 7-32 Roppongi, Minatoku Tokyo, Japan

505 Commission of the European Communities, DG XIII/A2, P.O. Box 1907, Luxembourg

506 International Association for Pattern Recognition (IAPR), Department of Physics Astronomy, University College London, Gower Street, London WC1E 6BT, England

507 International Cartographic Association (ICA), 24 Strickland Road, Mt. Pleasant, Perth 6153, Australia

508 International Translations Center (ITC), Doelenstraat 101, NL-2611 NS Delft, The Netherlands

509 Organization Européenne d'Etudes Photogrammétriques Expérimentales (OEEPE), 350 Boulevard 1945, P.O. Box 6, NL-7500 AA Enschede, The Netherlands

510 Secrétariat Général, African Remote Sensing Council (ARSC), P.O. Box 2335, Bamako, Republic of Mali

Algeria

511 Institut National de Cartographie, 123 rue de Tripoli, B.P. 32, Hussein-Dey, Algeria

Argentina

512 Comisión Nacional de Investigaciones Especiales (CNIE), Centro de Teleobservación, Av. del Liberator 1513, Vincente Lopez 1638, Buenos Aires, Argentina

Australia

513 Australian Mineral Foundation, 63 Conygham Street, Private Bag 97, Glenside, South Australia 5065, Australia
Tel. (08) 79 7821 Telex AA 87437
1. D. S. Crowe, Director
2. D. A. Tellis, Information Services Manager
Finance: Subscription and services
Purpose: To provide continuing education through intensive workshops involving training and refresher courses for personnel in the mining and petroleum industries and disciplines
Main Interests: Geological interpretation of aerial photographs and satellite images for exploration geosciences
Education and Training: Various courses associated with remote sensing and airphoto interpretation are conducted periodically, as well as a main course entitled 'Geological Interpretation of Aerial Photographs and Satellite Images'
Library: 5,000 volumes, 650 current journals
Publications: AESIS Quarterly, Earth Science and Related Information (monthly current awareness bulletin), *AESIS Special List No 12 'Remote Sensing and Photogeology 1976-1982'* (irregular)

514 Remote Sensing Association of Australia, Centre for Remote Sensing, University of New South Wales, P.O. Box 1, Kensington, New South Wales 2033, Australia
Tel. (02) 697 4183 Telex AA 26054
1. Assoc. Prof. B. C. Forster, President
Finance: Membership fees
Purpose: To support and sponsor the development of remote sensing through a professional organization
Main Interests: All areas of remote sensing
Education and Training: Host of the Australasian Remote Sensing Conference (every three years, most recent 1987)
Library: Uses the Central Library of the University of N.S.W, 1.3 million volumes, 16,500 journals, holding library for NASA publications
Publications: Remote Sensing Bulletin (annual)

515 Australian Landsat Station, 22-36 Oatley Court, P.O. Box 28, Belconnen, ACT 2616, Australia

516 Australian Mineral Foundation (AMF), Private Bag 97, Glenside, S.A. 5065, Australia

517 Australian Photogrammetric and Remote Sensing Society, P.O. Box 1020H, G.P.O. Melbourne,Victoria 3001, Australia

518 Commonwealth Scientific and Industrial Reasearch Organization (CSIRO), P.O. Box 225, Dickson, ACT 2602, Australia

519 Map Library, Dixon Wing, City Road, Institute Building, Sydney, N.S.W. 2006, Australia

520 NATMAP, Division of National Mapping, P.O. Box 31, Belconnen, ACT 2616, Australia

521 South Australian Department of Mines and Energy, P.O. Box 151, Eastwood, South Australia 5063, Australia

522 South Australian Institute of Technology, P.O. Box 1, Ingle Farm, South Australia 5098, Australia

523 University of New South Wales, School of Surveying, P.O. Box 1, Kensington, N.S.W. 2033, Australia

Austria

524 Austrian Space and Solar Agency - ASSA, Garnisongasse 7, A-1090 Vienna, Austria

525 Beckel Satellitenbilddaten, Marie-Louisen Strasse, A-4820 Bad Ischl, Austria

526 Universität für Bodenkultur, Geodesy, Photogrammetry and Remote Sensing Institute, Gregor Mandel-Strasse 33, A-1180 Vienna, Austria

Bangladesh

527 Space Research and Remote Sensing Organization (SPARSSO), Science and Techology Division, Cabinet Secretariat, House No. 23, Road No. 9/A, Dhanmondi R/A, Dacca 5, Bangladesh

Belgium

528 Belgian Science Policy Office, rue de la Science 8, B-1040 Brussels, Belgium
Tel. (02) 230 41 00 Telefax 230 59 12

1. Dr. P. Vanhaecke, Operational Director
2. L. Lofgren

Finance: Government

Purpose: To organize and co-ordinate research and development in remote sensing in Belgium

Main Interests: Application of second-generation remote sensing satellites in the fields of agriculture, forestry, spatial planning, cartography, earth sciences, marine sciences and natural resources in developing countries

Publications: A list of co-operating research laboratories and institutions

Other: The Belgian Science Policy Office has recently started a National Remote Sensing Programme involving seventeen national research laboratories and institutes. A central remote sensing unit for the treatment, archiving and distribution of remote sensing data is planned.

529 Société Belge de Photogrammétrie, de Télédétection et de Cartographie (SBPT), C.A.E. Tour Finances (Boîte No. 38), Boulevard de Jardin Botanique 50, B-1010 Brussels, Belgium
Tel. (02) 2103575
1. Andre Verduin, Director
2. Henri Van Olffen
Purpose: Co-ordination, promotion and dissemination of information on photogrammetry, remote sensing and cartography
Main Interests: Interpretation of satellite imagery for agriculture and cartography. Member of the International Society for Photogrammetry and Remote Sensing
Education and Training: Monthly conference held on photogrammetry, remote sensing or cartography
Publications: Bulletin de la Société Belge de Photogrammétrie, Télédétection et Cartographie (bi-annual)

530 Institut Géographique National (National Geographic Institute), Abbaye de Cambre 13, B-1050 Brussels, Belgium

Bolivia

531 Centro de Investigación y Aplicación de Sensores Remotos (CIASER), Casilla de correo 2729, La Paz, Bolivia

Brazil

532 Instituto de Pesquisas Espaciais, Departamento de Aplicações de Dudos de Satelite, P.O. Box 515, São José dos Campos - SP, São Paulo State 12225, Brazil
Tel. (0123) 22 9977 x250 Telex 1133530 INPE - BR
Finance: Data sales
Purpose: Commercial company interested in research and development
Main Interests: Multidisciplinary, visible, near infrared, thermal and microwave remote sensing
Education and Training: Short and middle-term courses for professionals, postgraduate studies, active technology transfer programme
Library: 28,000 volumes, 1,335 journals, 35,000 papers, various AV resources. Over 200,000 images (South America) accessible via computer searches
Publications: Annual report, publicity brochure (irregular)

533 Instituto de Geociências e Ciências Exatas-UNESP, Rua 10, Caixa Postal 178, Rio Claro, São Paulo 13500, Brazil
Tel. 0195 340122
1. Prof. A. Christofoletti
Finance: Government
Purpose: Education and research
Main Interests: Land use, geology, geomorphology, regional planning, natural vegetation and soils

Education and Training: Undergraduate remote sensing courses, postgraduate courses involving remote sensing. Three-month training courses leading to the certificate of Specialist. Annual Quantitative Geography Symposium and annual Physical Geography Symposium
Library: 30,000 volumes, 250 current periodicals
Publications: *Geografia* (bi-annual), *Geografia Teoretica* (bi-annual). Staff publish in these and other South American journals and conference proceedings; a publications list is available.

534 INPE-DGI, Caixa Postal 01, Cachoeira Paulista-CEP 12,630, São Paulo, Brazil

535 Instituto de Desenvolvimento de Pernambuco (CONDEPE), Departamento de Desenvolvimento de Systemas, Divisão de Informatica, Bibliotheca, rue Gervasio Peres, 399 Boa Vista, CP 1374, 5000 Recife, Brazil

536 Sensora, Avenida Sernambetiba, NR 4446, Rio de Janeiro CEP 22600, Brazil

537 South American Association of Remote Sensing, Sociedade Brasileira de Cartografia, Rua México 41, Sala 706, Centro Rio de Janeiro, Brazil 20 000

Burkina Faso

538 Centre Régional de Télédétection de Ouagadougou, B.P. 1762, Ouagadougou, Burkina Faso
Tel. 350 91/351 39 Telex CRETED 53 22 UV
1. Leon Okio, Director General
2. Sabaly Traore, Librarian
Finance: Government
Purpose: Regional remote sensing development in western and central Africa through training and user assistance services
Main Interests: Agronomy, forage, resources, forestry, hydrology, hydrogeology, monitoring and desertification, urban development, land use and soil improvement
Education and Training: Remote sensing application course for engineers and technicians, two per year in French and English, each of nine months' duration. Application of remote sensing to agricultural statistics for engineers, one per year, four months' duration. Application of remote sensing to hydrology and hydrogeology (a new course of three months' duration). Irregular seminars on various aspects of remote sensing
Library: 2,000 volumes, 30 current periodicals. Over 1,000 Landsat 1, 2 and 3 scenes (W. Africa), 300 aerial photographs (Burkina), image mosaics of the Ivory Coast, Niger, Mali and Burkina Faso
Publications: Annual report, technical reports, students' reports

Canada

539 Canada Centre for Remote Sensing, Department of Energy, Mines and Resources, 2464 Sheffield Road, Ottawa, Ontario K1A 0Y7, Canada

Tel. (613) 993 9900 Telex 053 3777
1. E. A. Godby, Director General
2. B. McGurrin, Head of the Technical Information Service
Finance: Government
Purpose: To co-ordinate remote sensing in Canada, to provide information on remote sensing and to perform research and development
Main Interests: Research and development in sensors, systems, image processing, applications and related sciences
Education and Training: Not an educational institution, but regular co-sponsor of the Canadian Symposium on Remote Sensing
Library: 2,050 volumes, 130 current periodicals. Microfiche index to Landsat imagery, filed images (Skylab, NOAA), airborne flight index maps
Publications: *Remote Sensing in Canada* (quarterly newsletter in English and French), *Canadian Advisory Committee on Remote Sensing* (annual minutes)

540 Ontario Centre for Remote Sensing, Ontario Ministry of Natural Resources, 880 Bay Street, 3rd Floor, Toronto, Ontario M5S 1Z8, Canada
Tel. (416) 965 8411 Telex 06219 701
1. Victor Zsilinszky, Director
Finance: Government
Purpose: Remote sensing research and development, trial applications, training
Main Interests: Development of applications of spaceborne and airborne remote sensing in forestry, geology, land cover mapping, peatland inventory, agricultural inventory and engineering studies
Education and Training: Short courses offered; photo-interpretation and remote sensing for Boreal Forest conditions; photo-interpretation and remote sensing for Great Lakes-St. Lawrence forest conditions; supplementary aerial photography. Practical short course in remote sensing. All courses are four- to five-day. Technology transfer programme with companies, government and universities
Library: 2,000 volumes, applications slide sets. Landsat of Ontario, aerial photography taken by OCRS for various projects
Publications: List available on request.

541 Air Photo Distribution Centre, Alberta Energy and Natural Resources, 2nd Floor West, North Tower, Petroleum Plaza, 9945 - 108th Street, Edmonton, Alberta T5K 2G6, Canada

542 Air Photo Sales and Information, Department of Mines, Resources and Environmental Management, 1007 Century Street, Winnipeg, Manitoba R3H 0W4, Canada

543 Air Photo Sales and Information, Energy, Mines and Resources, Regional Surveyor's Office, Bellanca Building, Room 314, Yellowknife, Northwest Territories X1A 2N5, Canada

544 Air Photo Sales and Information, Geological Survey of Canada, Institute of Sedimentary and Petroleum Geology, 3303 - 33rd Street N.W., Calgary, Alberta T2L 2A7, Canada

545 Air Photo Sales and Information, Maritime Resource Management Service, Box 310, Amherst, Nova Scotia B4H 3Z5, Canada

546 Air Photo Sales and Information, Ministry of the Environment, Parliament Buildings, Victoria, British Columbia V8V 1X5, Canada

547 Alberta Remote Sensing Centre, Oxbridge Place, 9820 - 106th Street, 11th Floor, Edmonton, Alberta T5K 2J6, Canada

548 Canada Centre for Remote Sensing (CCRS), User Assistance and Marketing Unit, 717 Belfast Road, Ottawa, Ontario K1A 0Y7, Canada

549 Canadian Institute of Surveying, Box 5378, Sta. F, Ottawa, Ontario K2C 3J1, Canada

550 Canadian Remote Sensing Society, 222 Somerset Street West, Suite No. 601, Ottawa, Ontario K2P 0J1, Canada

551 Canadian Remote Sensing Training Institute (CRSTI), P.O. Box 8321, Ottawa Terminal, Ottawa, Ontario K1G 3H8, Canada

552 Dalhousie University, Macdonald Science Library, 11500 Halifax, Nova Scotia B3H 4J3, Canada

553 Department of Development, 5th Floor, Atlantic Place, St. John's, Newfoundland A1C 5T7, Canada

554 Department of Forest Resources and Lands, Air Photo and Map Library, Hawley Building, Higgins Line, St. John's, Newfoundland A1C 5T7, Canada

555 Department of Renewable Resources, Government of the N.W.T., Yellowknife, Northwest Territories X1A 2L9, Canada

556 DIGIM, 1100 Blvd Dorchester West, Montreal, Quebec H3B 4P3, Canada

557 Energy, Mines and Resources, Regional Surveyor's Office, Air Photo Sales and Information, 200 Range Road, Room 208, Whitehorse, Yukon Y1A 3V1, Canada

558 Energy, Mines and Resources Canada, Research and Technology Sector, 580 Booth Street, Ottawa, Ontario K1A 0E4, Canada

559 Energy, Mines and Resources Canada, Surveys and Mapping Branch, 615 Booth Street, Ottawa, Ontario K1A 0E9, Canada

560 Manitoba Remote Sensing Centre, 1007 Century Street, Winnipeg, Manitoba R3H 0W4, Canada

561 Maritime Remote Sensing Committee, c/o Maritime Resource Management Service, P.O. Box 310, Amherst, Nova Scotia B4H 3Z5, Canada

562 Ministry of Forests Planning and Inventory Branch, 1450 Government Street, Victoria, British Columbia V8W 3E7, Canada

563 Ministry of Natural Resources, Ontario Centre for Remote Sensing, 880 Bay Street, 3rd Floor, Toronto, Ontario M5S 1Z8, Canada

564 National Air Photo Library (NAPL), 615 Booth Street, Ottawa, Ontario K1A 0E9, Canada

565 New Brunswick Department of the Environment, P.O. Box 6000, Fredericton, New Brunswick B3B 5H1, Canada

566 Nova Scotia Remote Sensing Committee, c/o Nova Scotia Land Survey Institute, P.O. Box 10, Lawrencetown, Nova Scotia B0S 1M0, Canada

567 Remote Sensing Centre, P.O. Box 1600, Charlottetown, Prince Edward Island C1A 7N3, Canada

568 Resource Planning Branch, Government of the Yukon, P.O. Box 2703, Whitehorse, Yukon Y1A 2C6, Canada

569 Saskatchewan Research Council, 30 Campus Drive, Saskatoon, Saskatchewan S7N 0X1, Canada

570 Service de Télédétection, Ministère des Terres et Forêts, 1995 Boul. Charest Ouest, Ste-Foy, Québec G1N 4H9, Canada

571 Université du Québec à Montréal, Service des Bibliothèques, 1225 rue St. Denis, P.O. Box 8889 Scc. A, Montreal, Québec H3C 3P3, Canada

572 Université Laval, Bibliothèque, Pavillon Bonefont, Québec G1K 7P4, Canada

573 University of British Columbia, Department of Geography, 1984 West Mall, Vancouver, British Columbia V6T 1W5, Canada

574 University of Ottawa, Library System, 65 Hastey Avenue, Ottawa, Ontario K1N 9A5, Canada

575 University of Toronto, Libraries, 130 St George Street, Toronto, Ontario M5S 1A5, Canada

Chile

576 Instituto Geográfico Militar, Sección Sensores Remotos, Nueva Santa Isabel No 1640, Santiago, Región Metropolitana, Chile
Tel. 6968221
1. Ingeniero Enrique López Silva, Director
Finance: Government

Purpose: Geography, photogrammetry and cartography
Main Interests: Cartography and remote sensing
Library: Fifty-six Landsat scenes, panchromatic aerial photography (whole country), infrared aerial photography (5% of the country)
Publications: *Boletín Informativo del Instituto Geográfico Militar* (irregular)

577 Instituto Nacional de Investigación de Recursos Naturales, Centro de Documentation, Manuel Montt 1164, Casilla 14995, Santiago, Chile

578 Servicio Aerofotogramétrico de la Fuerza Aerea (SAF), Casilla 67 correo los cerillos, Santiago, Chile

China

579 Scientific Research Institute of Surveying and Mapping, Department of Photogrammetry and Remote Sensing, No. 7 Yong Ding Road, Beijing, People's Republic of China
Tel. 81 2677 / 81 2158 Telex 3261
1. Zhang Xiao Rong, President of the Institute
Finance: Government
Purpose: Research into fundamental theories, methods and new techniques of surveying and mapping, including the cartographic applications of remotely sensed data. Solving the technical problems encountered in construction and surveying in China
Main Interests: Cartographic applications of remote sensing. Geographic information systems. Provision of an information service
Other: No further information supplied

580 Department of Photogrammetry and Remote Sensing, Wuhan Technical University of Surveying and Mapping, 23 Lo-Yu Road, Wuhan, Hubei Province, People's Republic of China
Tel. 875571 Telex 40170 WCTEL CN
1. Professor Zhu Xun Zhang, Head of Department
Finance: Government
Purpose: Teaching and research
Main Interests: Digital image processing and pattern recognition, geographic information systems and artificial intelligence. Air and space photogrammetry. Application of remote sensing in various fields of the national economy
Education and Training: Courses with a variable content of remote sensing and photogrammetry leading to BSc or MSc degrees. Occasional seminars on remote sensing, photogrammetry, cartography and related topics
Library: 1,438,000 volumes, 5,000 current periodicals, 430 microforms, slide sets and A/V materials for teaching purposes. Landsat MSS, Spacelab, SPOT and aerial imagery
Publications: *Journal of Wuhan Technical University of Surveying and Mapping* (irregular)

581 Academia Sinica, Landsat Ground Station, Beijing, People's Republic of China

582 China National Committee on Remote Sensing, State Scientific and Technological Commission of China, San Litti, Beijing, China

583 Space Science and Technology Center, Chinese Academy of Science, Beijing, People's Republic of China

Colombia

584 Centro Interamericano de Fotointerpretación, Carrera 30 No. 47-A-57, Apartado Aéreo 53754, Bogotá D.E., Cundinamarca 2, Colombia
Tel. 2680300 Telex 45656 DMOPT CO
1. M. Rivera, Director General
2. R. Liliana Osorio, Head of Information

Finance: Government and consultancy
Purpose: To implement plans, programmes, research and consultancy in the use of imagery taken of the Earth's surface that are directed towards the development of renewable and non-renewable resources
Main Interests: Digital image processing, visual interpretations, geographic information systems
Education and Training: Postgraduate courses in various application fields, short courses (six weeks' duration) on remote sensing in natural resource surveys and engineering; course brochure supplied upon request. Hosts the Colombian remote sensing symposium and holds irregular seminars on engineering geology and remote sensing applied to natural resources
Library: 5,000 volumes, 223 current periodicals, 200 manuscripts. Landsat of Atlantic and Pacific coasts
Publications: Revista CIAF (quarterly), *SENSOR Newsletter* (irregular), staff publications list available upon request
Other: CIAF is implementing a remote sensing programme with financial aid from the Dutch government, Interamerican Development Bank (IDB) and Organization of American States (OAS)

Cuba

585 Instituto Cubano de Geodesia y Cartografía, Loma y 39, Nuevo Vedado, Havana, Cuba

Cyprus

586 Cyprus Photogrammetric and Cartographic Association, Lands and Surveys Department, Archbishop Makarios III Avenue, P.O. Box 5598, Nicosia, Cyprus

Czechoslovakia

587 Czech Remote Sensing Centre, Czech Office of Geodesy and Cartography, Prague, Czechoslovakia

Denmark

588 Danish Society for Photogrammetry and Surveying, Aalborg Universitetscenter, Fibigerstraede 11, DK-9220 Aalborg Ost, Denmark

589 Electromagnetic Institute, DTH - Building 348, DK-2800 Lyngby, Denmark

590 Plancenter Fyn A/S, Overgade 32, DK-5000 Odense C, Denmark

Ecuador

591 Centro de Levantamientos Integrados de Recursos Naturales por Sensores Remotos (CLIRSEN), Subgerencia de Teledetección, Calle Paz y Miño s/n Edf. I.G.M., 4 to. piso/Casilla 8216, Quito, Ecuador
Tel. 542758 Telex 2775 CLIRSN ED
1. Eduardo Silva Maridueña, Executive Director
Finance: Government
Purpose: Formation of an inventory of renewable and non-renewable resources
Main Interests: All
Education and Training: Short courses and seminars are held on various aspects of remote sensing
Library: Slide sets and videos. Landsat 1, 2, 3, 4 and 5 of Ecuador. SPOT, GOES, NIMBUS, SIR-B and SLAR imagery
Publications: *Revista Teledetección* (annual report); publication list available on request

Egypt

592 Remote Sensing Center, 101 Kasr El Aini Street, Cairo, Egypt
Tel. 3557110 Telex 93069 ASRT UN
1. Prof. Dr. M. A. Abel Hady, Director of Remote Sensing Center
Finance: Government
Purpose: Research, training and projects dealing with the economic development of Egypt and surrounding countries
Main Interests: Remote sensing methodologies and applications in geology, agriculture, fisheries and land use
Education and Training: Short courses in various applications fields; occasional seminars
Library: Under organization
Publications: Staff publication list available on request

Ethiopia

593 Ethiopian Mapping Agency, Survey, Mapping and Remote Sensing Techniques Division, P.O. Box 597, Addis Ababa, Ethiopia
Tel. 44 84 45 Telex 21140 MAP ET
1. Mangesha Woldesemait, Head
2. Kassahun Sime, Information Officer
Finance: Government, and Food and Agriculture Organization

Purpose: To strengthen the remote sensing application and technological capability of the agency and extend its services to other governments. Training in remote sensing techniques. The design and implementation of various projects
Main Interests: Map revision, location, inventory and assessment of natural resources
Education and Training: Training given in co-operation with the FAO, full time for eight months, on: general concepts of remote sensing, four months' duration; application of remote sensing to hydrogeology and forestry, two months' duration each
Library: Small image mosaics, purchase of remotely sensed imagery in progress

Finland

594 Institute of Photogrammetry, Helsinki University of Technology, Ootakari 1, SF-02150 Espoo 15, Finland
Tel. 0 451 2523 Telex 125 161 HTKK SF
1. Prof. Dr. E. Kilpeä
Finance: Government
Purpose: Research and education
Main Interests: Digital image processing
Education and Training: Various training and lecture sessions as part of an MSc course
Library: No details available
Publications: *Photogrammetric Journal of Finland* (1-2 p.a.)

595 Geodeettinen Laitos (Geodetic Institute of Finland), Ilmalankatu 1A, SF-00240 Helsinki 24, Finland

596 National Board of Survey (NBS), Pasilan Virastukeskus, Opastinsilta, 12 Helsinki 52 00521, Finland

France

597 Institut Géographique National, 2 avenue Pasteur, F-94260 Saint-Mandé, France
Tel. 43 74 12 15
1. B. Galtier, Chief of the remote sensing division
Finance: Government
Purpose: Mapping, remote sensing and geodesy
Main Interests: Cartography
Education and Training: Professional teaching is given by the GDTA at its headquarters in Toulouse. It ranges from short courses to two- to three-year postgraduate studies
Library: 50,000 volumes, 100 current periodicals, 3,000 manuscripts, 1,500,000 maps. Over 300 MSS and TM scenes, thousands of aerial photographs, space maps

Publications: Annual report; *Bulletin d'Information* (newsletter, four to six per year)
Other: The Institut Géographique National is one of the founder members of the Groupement pour le Développement de la Télédétection Aérospatiale (GDTA - French Aerospace Remote Sensing Development Organization), the co-ordinating body in France for remote sensing activities

598 SPOT IMAGE, 16 bis avenue Edouard Belin, F-31030 Toulouse Cedex, France
Tel. 61 2731 31 Telex 532 079 F
1. A. Fontanel, Head Manager
Finance: Commercial company
Purpose: Distribution of SPOT satellite images
Main Interests: Production and distribution of earth observation images. All applications
Education and Training: Short courses on the SPOT system, duration two weeks, four per year (two in English, two in French). SPOT IMAGE regularly exhibits its products at international conferences
Library: Publicity films. Worldwide catalogue containing information on the images acquired by all SPOT receiving stations, the catalogue being available to all users through any conventional telecommunication link
Publications: Annual report, *SPOT newsletter* (bi-annual), various publicity materials (irregular), SPOT reference grid (references of SPOT scenes worldwide)

599 Prospace, 2 place Maurice Quentin, F-75039 Paris Cedex 1, France
Tel. (33-1) 45 08 77 70 Telex 214674 F
1. A. M. Gaubert, Chief executive
2. E. M. Cerf-Mayer, Marketing executive
Finance: Internal
Purpose: The promotion of French space industries' activities, including remote sensing
Main Interests: SPOT
Education and Training: A regular information service is supplied and can be requested to provide additional information and contact with members of the Prospace group. All services are free of charge
Publications: *French remote sensing activities, services and products* (annual catalogue), *News from Prospace* (three per year), Prospace catalogue, several volumes (irregular)

600 Bureau de Richerches Géologiques et Minières (BRGM), 191 rue de Vaugirard, F-75737 Paris 15, France

601 Bureau pour le Développement de la Production Agricole (BDPA), 27 rue Louis Vicat Immeuble (Le Bearn), F-75015 Paris, France

602 Centre National de la Recherche Scientifique (CNRS), 26 rue Boyer, F-75971 Paris 20, France

603 Centre National d'Études Spatiales - CNES (National Centre for Space Studies), 129 rue de l'Université, F-75007 Paris, France

604 Comité de la Recherche Spatiale - COSPAR (Committee on Space Research), 51 boulevard de Montmorency, F-75016 Paris, France

605 Groupement pour le Développement de la Télédétection Aérospatiale - Centre Spatiale de Toulouse, 18 avenue Edouard Belin, F-31055 Toulouse, France

606 Institut Français du Pétrole (IFP), 1-4 avenue du Bois Préau, F-92506 Rueil Malmaison, France

607 Société Française de Photogrammétrie et de Télédétection, 2 avenue Pasteur, F-94160 Saint Mandé, France

The Gambia

608 Survey Department of the Gambia, Cotton Street, Banjul, The Gambia

German Federal Republic

609 Deutsche Forschungs- und Versuchsanstalt für Luft- und Raumfahrt WT-DA-FE (DFVLR), D-8031 Oberpfaffenhofen, Federal Republic of Germany
Tel. 08153/28 885 Telex 526 419 DVLOP D
1. Dr. Rudolf Winter, Head of remote sensing section
Finance: Non-profit, funded by federal ministries
Purpose: Aerospace research
Main Interests: Mapping of renewable resources and forest disease, exploration of mineral deposits and environmental monitoring
Education and Training: Annual international training courses on remote sensing of renewable resources, jointly organized for four weeks in Germany with the Food and Agriculture Organization. Yearly one-week courses on remote sensing and image processing. Annual user seminars and an international forum for the German remote sensing data centre
Library: 65,000 volumes, 310 current periodicals, scientific reports, slide sets and videos. Landsat, Seasat, NOAA, Meteosat and Bendix scanner imagery
Publications: Zeitschrift für Flugwissenschaften und Weltraumforschung (bi-monthly). DFVLR annual report. Publication lists are available upon request

610 Institut für Angewandte Geodäsie (IFAG), Abteilung Photogrammetrische Forschung, Richard-Strauss-Allee 11, D-6000 Frankfurt am Main 70, Federal Republic of Germany
Tel. (069) 63 33-1 Telex 04 13 592 IFAG D
1. Prof. Dr.-Ing Heinz Schmidt-Falkenburg
Finance: Government

Purpose: Scientific research work in all fields of geodesy, cartography and photogrammetry and in the adaption of the research results for practical application
Main Interests: Evaluation of data in passive and active image recording systems on photogrammetry and remote sensing. Participation in the space programmes of the Federal Republic of Germany and the ESA on Earth reconnaissance and measurement. Image processing and correction of aerial photograph and satellite data
Education and Training: Advanced training of specialists and executives from developing countries and United Nations and government scholarship holders
Library: 58,000 volumes, 500 current periodicals, translation bureau for publications, slide sets. Landsat and NOAA imagery
Publications: Nachrichten aus dem Karten- und Vermessungswesen (irregular). Satellite maps and mosaics. Publication and price lists available on request

611 Carl Zeiss, Postfach 1369/1380, D-7082 Oberkochen, Federal Republic of Germany

612 Federal Institute for Geosciences and Natural Resources, Geoscience Literature Information Service, Stillweg 2, Postfach 510 153, D-3000 Hannover 51, Federal Republic of Germany

613 German Society for Photogrammetry and Remote Sensing, c/o Institute for Photogrammetry, Nienburger Strasse 1, D-3000 Hannover, Federal Republic of Germany

German Democratic Republic

614 Jenoptik Jena GmbH, Carl-Zeiss-Strasse 2, 69 Jena, German Democratic Republic

Greece

615 Greek Society of Photogrammetry, 61 Arachovis Street, Athens, Greece

Guinea

616 Institut Géographique National de Guinea, B.P. 1151, Conakry, Guinea

Hong Kong

617 Geocarto International Centre, GPO Box 4122, Hong Kong

Hungary

618 Földmérési Intézet (FÖMI), Remote Sensing Centre, P.O. Box 546, Guszev u. 19, 1051 Budapest 5, Hungary
Tel. 361 636 669 Telex 22 4964
1. S. Zsamboki, Head of Remote Sensing Centre
Finance: Government bodies
Purpose: To provide services and education, allied with research and development in all methods and applications of remote sensing. Co-operation in Intercosmos, ESA Earthnet and distributor for SPOT data in Hungary

Main Interests: Optical near-infrared remote sensing, quantitative analysis of digital satellite data, analog processing and evaluation of airborne imagery
Education and Training: Various short courses, conferences and seminars on a range of remote sensing research and application topics
Library: 30,000 volumes, 100 current periodicals, slide sets and videos. Partly computerized browse archive of aerospace imagery. Eight hundred Landsat 1, 2, 3, 4 and 5 scenes of Hungary, 25,000 aerial photographs
Publications: Távérzé Kelési Korlevél (irregular)

619 Cartographia, P.O. Box 132, 1443 Budapest, Hungary

India

620 National Remote Sensing Agency (NRSA), Department of Space, Balangar, Hyderabad 500 037, Andhra Pradesh, India
Tel. 0842 262572/263360 Telex 0155 6522
1. Laxmana Bulusu, Director
2. Srinivasa Adiga, Technical secretary
Finance: Government
Purpose: Satellite data acquisition, product generation and dissemination. Flight operation to provide aerial imagery. Application of remote sensing to resource surveys and environmental monitoring. Training
Main Interests: Remote sensing data correction. Image processing. Remotely sensed data applications
Education and Training: Diploma courses in various disciplines, ten months' duration, offered at the Indian Institute of Remote Sensing. Short certificate courses in aerial photography and satellite imagery applications
Library: 3,433 volumes, 231 current periodicals, 500 technical reports. Over 75 satellite and aerial mosaics, 4,050 CCTs of Landsat 2, 3, 4 and 5 data, 110 CCTs aerial scanner data, 200,000 aerial photographs
Publications: Annual report; report on technical activities (annual); technical reports. List available on request

621 Centre of Studies in Resources Engineering (CSRE), Indian Institute of Technology (IIT), Powai, Bombay 400 076, India

622 Indian Institute of Remote Sensing (IIRS), National Remote Sensing Agency (NRSA), No. 4 Kalidas Road, P.O. Box 135, Dehra Dun (UP) 248 001, Uttar Pradesh, India

623 Indian Space Research Organization (ISRO), Cauvery, Bahvan, District Office Road, Bangalore 560 009, India

624 Space Applications Centre (SAC), Ahmedabad 380 083, Gujarat, India

625 Survey of India, P.O. Box 37, Hathibarkala, Dehra Dun, India

Indonesia

626 Badan Koordinasi Survey dan Pemetaan Nasional (National Co-ordination Agency for Surveys and Mapping), Jalan Raya Bogor, Km 46, Cibinong, Bogor, Indonesia

627 Center for Remote Sensing Image Interpretation and Integrated Surveys (PUSPICS), Faculty of Geography, Gadjah Mada University, Bulaksumur, Yogyakarta, Indonesia

628 Indonesian National Institute of Aeronautics and Space (LAPAN), JL Pemuda Persil No. 1, P.O. Box 3048, Jakarta, Indonesia

Iran

629 Remote Sensing Centre of Iran, Planning and Budget Organization, No. 80, West Sepand Avenue, Tehran, Iran

Iraq

630 Remote Sensing Department, Space and Astronomy Research Center, Scientific Research Council, P.O. Box 2441, Jadriyia, Baghdad, Iraq

Tel. 9641 7765116 Telex 213976 SRC

1. Q. A. Abdullah, Head of Remote Sensing Department

Finance: Government

Purpose: Remote sensing research and developement

Main Interests: Remote sensing applications in agriculture and water resources. Digital image processing

Education and Training: Preliminary level training course in geological, hydrological, mapping and land use applications of remote sensing, two weeks' duration. Host of First National Symposium on Remote Sensing, various other seminars at irregular intervals

Library: 12,000 volumes, 1,098 current periodicals. Slide sets and videos

Ireland

631 ERA, Remote Sensing Geological and Environmental Services, Environmental Resources Analysis Ltd., 187 Pearse Street, Dublin 2, Ireland

632 Irish Society of Surveying and Photogrammetry, Engineering School, Trinity College, Dublin 2, Ireland

633 National Board for Science and Technology, Shelbourne House, Shelbourne Road, Dublin 4, Ireland

Israel

634 Department of Surveys, P.O. Box 14171, Tel Aviv 611141, Israel

635 Interdisciplinary Centre for Technology Analysis and Forecasting, Tel Aviv University, RAMAT AVIV, 69 978 Tel Aviv, Israel

Italy

636 Consiglio Nazionale delle Richerche (CNR), Piazzale Aldo Moro 7, I-00100 Rome, Italy

637 Istituto Geografico Militare, Via C. Battisti 10, I-50100 Florence, Italy

638 Ministero delle Finanze, Largo Leopardi No 5, I-00185 Rome, Italy

639 Società Italiana di Fotogrammetria e Topografia, Piazzale R. Morandi 2, I-20121 Milan, Italy

640 Telespazio S.p.A., Dipartimento Commercial, Via Alberto Bergamini 50, I-00159 Rome, Italy

Japan

641 Remote Sensing Technology Center of Japan, Uni-Roppongi Building, 7-15-17, Roppongi, Minato-ku, Tokyo 106, Japan
Tel. 03 403 1761 Telex 02426780 RESTEC J
1. Keiji Maruo, Managing Director
Finance: Commercial foundation
Purpose: Research and development in remote sensing in order to contribute to economic development, social welfare and environmental protection
Main Interests: Co-operation in education, training and popularization of remote sensing. Research and investigation of remote sensing technology and the collection and distribution of remote sensing data
Education and Training: Training courses in remote sensing, digital image analysis and processing, four- to eight-week duration. Occasional seminars on various remote sensing topics in collaboration with other groups
Publications: Data distribution guides (irregular), *RESTEC Newsletter* (bi-monthly), *RESTEC Journal* (quarterly)

642 Japan Society of Photogrammetry and Remote Sensing (JSPRS), Daiichi Honan Building, 601 2-8-17, Minami-ikebukuro, Toshima-ku, Tokyo, Japan

643 National Space Development Agency, 2-4-1, Hamamatsu-Cho, Minato-Ku, Tokyo 105, Japan

Jordan

644 Jordan National Geographic Center, Remote Sensing Center, P.O. Box 20214, Amman, Jordan
Tel. 845 188 Telex 22472
1. Brigadier General Rafat Majali, Director General
Finance: Government
Purpose: To meet Jordan's needs in remote sensing
Main Interests: Cartography, remote sensing, geodesy, training
Education and Training: Beginners'-level courses in the principles and applications of remote sensing, twenty days' duration

Kenya

645 Regional Centre for Services in Surveying, Mapping and Remote Sensing, P.O. Box 18118, Nairobi, Kenya
Tel. 803320/9 Telex 25285 REGSURVEYS
1. Dr. H. M. Hassan, Director of remote sensing
Finance: Intergovernmental
Purpose: To provide technical remote sensing services to earth resource scientists and decision makers from the eastern, central and southern African countries
Main Interests: Multidisciplinary
Education and Training: Discipline-specific training courses for established professionals; three to four weeks of theoretical work are complemented by an extended field project lasting up to six months. Annual information seminar, irregular seminars in other countries served by the Centre
Library: Small, slide sets and videos. Worldwide film positive collection of Landsat 1 and 2 images (1972-77), cloud-free MSS Band 5 for eastern, central and southern Africa (1972-78), all low cloud cover RBV from Landsat 3, various special products
Publications: *Earth Resources Mapping in Africa* (quarterly), various publicity materials (irregular)

Korea

646 Korea Advanced Institute of Science and Technology, P.O. Box 131, Dong Dae Mun, Seoul, Korea

Madagascar

647 Institut National de Géodésie et Cartographie, P.O. Box 323, 101 Antananarivo, Madagascar
Tel: 229-35
1. Solonavalona Andriamihaja, Engineer and Chief of Geography
Finance: Government and commercial
Purpose: Taking aerial photographs, surveying, mapping and printing
Main Interests: The up-date of maps and establishment of new thematic maps
Library: 3,000 volumes, 40 current periodicals. Six hundred image mosaics at a variety of scales, 7 Landsat scenes of south-west Madagascar, approximately 60,000 panchromatic aerial photographs of the whole country, and 4,000 historical photographs at a variety of scales

Malaysia

648 National Remote Sensing Committee (NRSC), Directorate of National Mapping, Kuala Lumpur, Malaysia

649 Terra Control Technologies, Sdn. Bhd, Godown 3, Banguman Nupro, Jalan Brickfield, 50470 Kuala Lumpur, Malaysia

Malawi

650 Geoservices Ltd., P.O. Box 30305, Lilongwe 3, Malawi

Mauritius

651 Service Topographique et de Cartographie, B.P. 237, Nouakchott, Mauritius

Republic of Mali

652 Direction Nationale de la Cartographie et de Topographie, B.P. 240, Bamako, Republic of Mali

Mexico

653 Centro Científico de América Latina-IBM (IBM of Mexico Scientific Center), Ruben Dario 55, Col. Chapultepec Polanco, CP. 11560, Mexico D.F.
Tel: 250-9011
1. Jaime Villanueva G, Scientific Centres Manager
Finance: Commercial
Purpose: IBM scientific centre to help investigations in Mexico
Main Interests: Agriculture, vegetation, water, soil, geology, land use, cartography and geothermy
Education and Training: No training is given in remote sensing, but the centre does promote conferences on remote sensing topics
Library: A library service is provided, but no details are available. Twenty-seven CCT's of Landsat data and corresponding cartographic maps are available for different areas of Mexico
Publications: A list of staff publications (mainly in Spanish) is available on request

654 Instituto Nacional de Estadística Geográfica e Informatica (INEGI), San Antonio Abad 124, Mexico 8, D.F., Mexico

655 Sociedad Mexicana de Fotogrametría, Fotointerpretación y Geodesia, Apartado Postal 25-447, Mexico 13, D.F., Mexico

Mozambique

656 Direcção Nacional de Geografia e Cadastro (National Directorate for Geography and Land Surveying), Josina Machel No 537, P.O. Box 288, Maputo, Mozambique

Nepal

657 National Remote Sensing Center (NRSC), P.O. Box 3103, Kathmandu, Nepal

The Netherlands

658 National Aerospace Laboratory NLR, Anthony Fokkerweg 2, NL-1059 CM Amsterdam, P.O. Box 90502, NL-1006 BM Amsterdam, The Netherlands
Tel. 20 5113 113 Telex 11118 MLRAA ML
1. Dr. N. J. J. Bunnik, Head Remote Sensing Unit

2. C. W. De Jong, Chief Librarian
Finance: Contract/research grant
Purpose: Aerospace research and development
Main Interests: Remote sensing technology development. Airborne remote sensing. Digital data processing, data distribution to users, contract studies, research campaigns
Library: Images from Landsat and the NLR laboratory aircraft
Publications: Annual report

659 Nationaal Lucht- en Ruimtevaartlaboratorium (NLR), P.O. Box 90502, NL-1006 BM Amsterdam, The Netherlands

660 Netherlands Remote Sensing Board (BCRS), Survey Department, Ministry of Transport and Public Works, P.O. Box 5023, NL-2600 GA Delft, The Netherlands

New Zealand

661 Division of Information Technology, Department of Scientific and Industrial Research, Private Bag, Lower Hutt, New Zealand
Tel. 04 666919 Telex NZ 3814 PHYSICS
1. P. J. Ellis, Director
2. S. E. Bellis, Scientist
Finance: Government
Purpose: To run the DSIR's national computer network, remote sensing and image processing
Main Interests: Research into the uses of satellite and aircraft remotely sensed data. Provision of products and services related to remote sensing
Education and Training: No formal training given; however, guest workers are appointed for varying time periods. Occasional seminars on various themes in remote sensing
Library: The division uses the DSIR's central library which is on site and has 580,000 volumes, 16,500 current periodicals, 25,000 government documents, 134,000 standard specifications, 20,000 maps, 110,000 microforms, slide sets. Satellite mosaics, 200 Landsat scenes (New Zealand, Antarctica, Samoa), 40 airborne scanner swaths (New Zealand), miscellaneous others
Publications: DSIR Remote Sensing Newsletter (annual); remote sensing publication list available

662 Geology Library, University of Auckland, 5 Symonds Street, Private Bag, Auckland 1, New Zealand

663 Remote Sensing Group, Soil Conservation Service (SCS), Hydrology Centre, Ministry of Works and Development, P.O. Box 1479, Christchurch, New Zealand

664 Remote Sensing Section, Physics and Engineering Laboratory (PEL), Department of Scientific and Industrial Research (DSIR), Private Bag, Lower Hutt, New Zealand

Nicaragua

665 Dirección de Geodesia y Cartografía (Geodetic and Cartographic Survey), 2110 Managua, Nicaragua

Nigeria

666 ECA Regional Centre for Training in Aerial Surveys (RECTAS), Department of Photo-interpretation and Remote Sensing, Department of Photogrammetry, PMB 5545, Ile-Ife, Oyo State, Nigeria

Tel. 036 2225 Telex. 2050 RECTAS

1. O. O. Ayeni, Director

Finance: Government and Member States

Purpose: Training and research in the applications of photo-interpretation and remote sensing to resources surveys and environmental management

Main Interests: Digital mapping, applications to resources surveys and environmental management

Education and Training: Short-term workshops in applications to agriculture, forestry, water resources, topographic and land-use map revision. Technicians' and technologists' course

Library: Small library. Satellite images (various transparencies, diapositives and black-and-white and colour paper prints), photographic images (film diapositives and paper prints)

Publications: A newsletter has just been launched, but no details are available as yet

667 Danz Surveys and Consultants, 24 Oyekan Road, Lagos, Nigeria

Norway

668 Fjellanger Wideroe A/S, P.O. Box 2916, N-7001 Trondheim, Norway

669 Tromso Telemetry Station, P.O. Box 387, 9001 Tromso, Norway

Pakistan

670 Pakistan Space and Upper Atmosphere Research Commission (SUPARCO), Remote Sensing Applications Centre (RESACENT), P.O. Box 3125, Karachi-29, Pakistan

Tel: 461151 Telex 25720 SPACE PAK

1. Hasan Zafrul, Head of the Remote Sensing Applications Centre
2. Muhammad Safeer Rana, Principal Scientific Officer

Finance: Government

Purpose: The promotion and co-ordination of remote sensing technology, applications and activities in Pakistan

Main Interests: Remote sensing technology development and thematic applications in the areas of agriculture, water resources, geology and mapping
Education and Training: Short training courses on remote sensing applications are arranged bi-annually for the national users' organization. The course duration is one week and it is offered to science graduates involved in resource surveying. Certificates of participation are awarded to the trainees on the completion of the course
Library: 3,200 volumes, 70 current periodicals, slide sets and videos. Over 2,000 scenes and 200 CCT's of Pakistan from 1972 onward, 59 SIR-A swaths and selected Metric Camera and NOAA/TIROS images
Publications: Space Horizons (quarterly), *SUPARCO Times*. A list of staff publications and reports is available on request

Panama

671 Defense Mapping Agency (DMA), Inter-American Geodetic Survey, Cartographic School, Drawer 936, Fort Clayton, Panama

Peru

672 Oficina Nacional de Evaluación de Recursos Naturales - ONERN (National Office for the Evaluation of Natural Resources), 355 calle 17, Urb El, Palomar-San Isidro, Lima, Peru

The Philippines

673 Natural Resources Management Center (NRMC), 8th Floor, P.O. Box AC, Quezon Avenue, Quezon City 493, The Philippines

Poland

674 University of Mining and Metallurgy, Photogrammetric Section, Al. Mickiewicza 30, 30-059 Krakow, Poland
Tel. 33 81 00 Telex 0322203 PL or 0325343 PL
1. Prof. Dr. Zbigniew Sitek, Head of photogrammetric research
2. Barbara Novak, Information officer

Finance: University budget
Purpose: Education and training
Main Interests: Digital image processing, rectification and classification
Education and Training: Various one-semester university courses as part of BSc degree. All-Polish symposium held every three years (the next is due in 1987) in collaboration with the Polish Society for Photogrammetry and Remote Sensing
Library: 15,000 volumes, 85 current periodicals, 2,000 reports and other media. Landsat and airborne imagery
Publications: Staff publications list available on request

675 Geokart, 2/4 rue Jasna, Warsaw 00-950, Poland

676 Polish Remote Sensing Centre, Institute of Geodesy and Cartography, Ul. Jasna 2/4, 00-950 Warsaw, Poland

677 Polska Akademia Nauk, Komitet Geodezji, Panstwowe Wydawnictwo Naukowe, Ul. Miodowa 10, 00-251 Warsaw, Poland

Portugal

678 Associação Portuguesa de Fotogrametria e Detecçao Remota, Av. Ilha da Madeira, 22-2oDto, 1400 Lisbon, Portugal

679 Geometral, Ave. Cons. Barjona de Freitas, No 20-A, 1500 Lisbon, Portugal

680 Instituto Geográfico e Cadastral (Geographic and Cadastral Institute), Praça da Estrela, 1200 Lisbon, Portugal

Romania

681 Romanian Committee of Photogrammetry and Remote Sensing, 79662 B-dul Expozitiei nr.1A, Sector 1, Bucharest, Romania

Saudi Arabia

682 King Abdulaziz City for Science and Technology, P.O. Box 6086, Riyadh, Saudi Arabia

Sénégal

683 Service Géographique de Sénégal, B.P. 740, 14 rue Victor Hugo, Dakar, Sénégal

Republic of South Africa

684 Satellite Remote Sensing Centre of the National Institute for Telecommunications Research, Council for Scientific and Industrial Research, P.O. Box 3718, Johannesburg 2000, Transvaal, Republic of South Africa

Tel. 011 642 4693 Telex 3-21005

1. W. J. Botha

Finance: Government/data sales

Purpose: Acquisition of data concerning the Earth's surface and atmosphere from satellites for dissemination to the user community

Main Interests: Acquisition and archival of remotely sensed data

Education and Training: No formal training given; image processing specialist and system made available to user community at nominal cost

Library: Over 38,000 Landsat 1, 2, 3, 4 and 5 MSS scenes (southern Africa)

Publications: Satellite Remote Sensing Centre Newsletter (quarterly), Landsat catalogues and publicity material (irregular)

685 South African Society for Photogrammetry, Remote Sensing and Cartography, P.O. Box 69, Newlands 7725, Republic of South Africa

Spain

686 CONIE, Pintor Rosales 34, Madrid 8, Spain

687 Instituto Geográfico Nacional (IGN), General Ibanez de Ibero, Apartado Aereo 3, Madrid 3, Spain

Sri Lanka

688 Arthur C. Clarke Centre for Modern Technologies, Katubedda, Moratuwa, Sri Lanka

689 Centre for Remote Sensing, Survey Department, P.O. Box 506, Colombo, Sri Lanka

Sudan

690 Sudan Society of Photogrammetry, Ministry of Defence, Survey Department, P.O. Box 306, Khartoum, Sudan

Suriname

691 Central Bureau for Aerial Mapping, Dr Sophie Redmondstraat 131, P.O. Box 971, Paramaribo, Suriname

Sweden

692 Swedish Space Corporation, Albygatan 107, S-171 54 Solna, Sweden
Tel. 46 87 33 6200 Telex 17128 SPACECO
1. H. Kihlberg, Director of the remote sensing division
2. S. Olovsson, Marketing Programmes Manager

Finance: Commerce/government
Purpose: Professional responsibility for the implementation of Sweden's space and remote sensing programmes
Main Interests: Receiving station for Landsat MSS, TM and SPOT. Image processing, geographic information systems, value added products. Consultancy and all fields of application
Education and Training: Short courses in digital image processing and geographic information systems. SPOT and geographic information system seminars
Library: 1,500 volumes, slide sets and videos. Landsat MSS, TM and SPOT coverage (worldwide)
Publications: *Swedish Space Corporation Newsletter* (quarterly), press releases (irregular)

693 ESRANGE, Fack, S-981 01 Kiruna, Sweden

694 Satimage, P.O. Box 816, 28 Kiruna S-981, Sweden

Switzerland

695 Swiss Meteorological Institute, Krahbuhlstrasse 58, CH-8044 Zurich, Switzerland
1. Dr. sc. nat. A. Praget, Chief

Finance: Government
Purpose: Operational and research application of satellite data to meteorology
Main Interests: General meteorological circulation, influence of the Alps on air circulation
Education and Training: Irregular conferences and seminars

Library: 4,170 volumes, 630 current periodicals, 520 sets of climatological data, DOCSAT - a leaflet system publication, especially containing references to the use of satellite data in meteorology. Meteosat imagery, daily for the last three years, slide sets and videos
Publications: Annual report. Irregular publications on DOCSAT. Publication list available on request

696 Bundesamt für Landestopographie, Seftigenstr. 264, CH-3084 Wabern, Switzerland

Syria

697 Military Survey Department, P.O. Box 3094, Damascus, Syria

698 National Remote Sensing Center (NRSC), P.O. Box 12586, Damascus, Syria

Taiwan

699 Center for Space and Remote Sensing Research, National Central University, Chung-Li, Taiwan 320

Thailand

700 Thailand Remote Sensing Centre, National Research Council, Ministry of Science, Technology and Energy, 196 Phahonythin Road, Bangkok 10900, Thailand
Tel. 579 0116 7 Telex 82213 NARECOUTH
1. S. Vibulsresth, Director
2. S. Suwan-Arpa, Information Officer
Finance: Government
Purpose: Remote sensing data acquisition, processing, applications and training
Main Interests: Landsat data reception and distribution. Forestry and land use
Education and Training: Annual remote sensing training course leading to a general-purpose certificate, one month's duration. Digital image processing course, certificated, one month's duration. Both in the Thai language
Library: 126,500 volumes, 850 current periodicals, 3,000 theses, 660,500 government documents, 134,000 standard specifications, slide sets. Over 34,000 Landsat scenes of Nepal, Sri Lanka, Tibet, China, Burma, Bangladesh, Indo-China, Philippines, Taiwan, Indonesia, Malaysia and Thailand. Aerial photographs
Publications: Annual report, *TRSC Newsletter* (quarterly)

701 Asian Regional Remote Sensing Training Centre (AARSTC), Asian Institute of Technology (AIT), P.O. Box 2754, Bangkok 10501, Thailand

Tunisia

702 Office de Topographie et Cartographie, 13 rue de Jordanie, Tunisia

United Kingdom

703 National Remote Sensing Centre (NRSC), R190 Building, Space Department, Royal Aircraft Establishment (RAE), Farnborough, Hampshire GU14 6TD, England

1. M. J. Hammond
2. L. Newbold

Finance: British National Space Centre

Purpose: To promote and foster the development of remote sensing, with particular emphasis on applications

Main Interests: National user services, research and development into image processing analysis and interpretation techniques, applications development, education and training

Education and Training: Beginner and advanced-level image processing workshops at the NRSC regional centre at Silsoe College

Library: 1,500 volumes, 117 journals and newsletters, online access to the ESA LEDA file, slide sets and videos. Landsat and SPOT (all U.K. and some global), U.K. and Ireland Landsat quick-look black-and-white prints, colour prints of some of the scenes held on the NRSC digital tape archive, microfiche catalogues of Seasat, HCMM and Metric Camera

Publications: National Remote Sensing Centre Newsletter (quarterly), *Current Awareness Bulletin* (monthly), *NRSC Data Users' Guide* (periodic updates). A series of 30 fact sheets that detail satellite/sensor characteristics and applications

704 Remote Sensing Unit, Department of Civil Engineering, Aston University, Aston Triangle, Birmingham B4 7ET, England

Tel. 021 359 3611 Telex 336997

1. Dr. W. G. Collins, Reader in remote sensing
2. E. C. Hyatt, Information specialist

Finance: Academic

Purpose: Academic research, education and training

Main interests: Natural resources inventories, environmental sciences and planning, earth sciences, information consultancy

Education and Training: PhD and MPhil research, specialist short courses from one week to one year (tailored to individual needs)

Library: Small specialist library, slide sets and videos. MSS, SIR-A, ATM imagery (30 scenes, U.K., Africa, India, U.S.A.), several thousand aerial photographs (various U.K. locations)

Publications: Remote Sensing Unit Brochure (irregular); a publications list is available on request

705 ADAS Aerial Photography Unit, Ministry of Agriculture, Fisheries and Food (MAFF), Block B, Brookland Avenue, Cambridge CB2 2DR, England

706 Air Photographs Unit, Royal Commission on the Historical Monuments of England (RCHME), National Monuments Record, Fortress House, 23 Savile Row, London W1X 1AB, England

707 Air Photographs Unit, Royal Commission on the Historical Monuments of England (RCHME), Room W210, Government Buildings, Bromyard Avenue, Acton, London W3 7AY, England

708 Air Photography Unit, Scottish Development Department (SDD), New St. Andrews House, Edinburgh EH1 3DH, Scotland

709 Aslib, Information House, 26/27 Boswell Street, London WC1N 3JZ, England

710 British Association of Remote Sensing Companies (BARSC), c/o Lansing Bagnall Building, Edenbridge, Kent TN8 6HS, England

711 British Interplanetary Society, 27/29 Lambeth Road, London SW8 1SZ, England

712 British Library Bibliographic Services Division (BLBSD), 2 Sheraton Street, London W1V 4BH, England

713 British Library Document Supply Centre (BLDSC), Boston Spa, Wetherby, West Yorkshire LS23 7BQ, England

714 British Library Map Room, Great Russell Street, London WC1B 3DG, England

715 British National Space Centre, Millbank Tower, Millbank, London SW1P 4QU, England

716 Carnegie Laboratory of Physics, University of Dundee, Dundee DD1 4HN, Scotland

717 Central Register of Air Photography for Wales, Welsh Office, Cathays Park, Cardiff CF1 3NQ, Wales

718 Centre for Remote Sensing, Imperial College of Science and Technology, Department of Physics, Prince Consort Road, London SW7 2AZ, England

719 Clyde Surveys Ltd., Environment and Resources Consultancy (ERC), Reform Road, Maidenhead SL6 8BU, England

720 Commonwealth Association of Surveying and Land Economy (CASLE), 12 George Street, Parliament Square, London SW1P 3AD, England

721 Eurimage, Hunting Surveys and Consultants Ltd., Elstree Way, Borehamwood, Hertfordshire WD6 1SB, England

722 Focal Point Audio Visual Ltd., 252 Copnor Road, Portsmouth PO3 5EE, England

723 General Technology Systems (GTS) Ltd., Forge House, 20 Market Place, Brentford, Middlesex TW8 8EQ, England

724 Geo Abstracts Ltd., Regency House, 34 Duke Street, Norwich NR3 3AP, England

725 Hunting Technical Services Ltd., Elstree Way, Borehamwood, Hertfordshire WD6 1SB, England

726 The Institution of Electrical Engineers (IEE), Station House, Nightingale Road, Hitchin, Hertfordshire SG5 1RJ, England

727 McCarta Ltd., 122 Kings Cross Road, London WC1X 9DS, England

728 Microinfo Ltd., P.O. Box 3, Alton, Hampshire GU34 1BA, England

729 Nigel Press Associates Ltd, Edenbridge, Kent TN8 6HS, England

730 Ordnance Survey of Northern Ireland (OSNI), Department of the Environment (NI), Colby House, Stranmills Court, Belfast BT9 5BJ, Northern Ireland

731 Ordnance Survey (OS), Romsey Road, Maybush, Southampton SO9 4DH, England

732 Photogrammetric Society, Department of Photogrammetry and Surveying, University College London, Gower Street, London WC1E 6BT, England

733 Remote Sensing Society, Department of Geography, University of Nottingham, University Park, Nottingham NG7 2RD, England

734 Remote Sensing Unit, Department of Agricultural Engineering, Silsoe College, Silsoe, Bedford MK45 4DT, England

735 Resource Planning Group, Land and Water Service (ADAS), Ministry of Agriculture, Fisheries and Food (MAFF), Great Westminster House, Horseferry Road, London SW1 2AP, England

736 Royal Meteorological Society, James Glaisher House, Grenville Place, Bracknell, Berkshire RG12 1BX, England

737 Space Frontiers Ltd., 30 Fifth Avenue, Havant, Hampshire PO9 2PL, England

738 Technical Indexes Ltd., Willoughby Road, Bracknell, Berkshire RG12 4DW, England

739 University of Keele, Department of Geography, Keele, Staffordshire ST5 5BG, England

United States of America

740 American Society for Photogrammetry and Remote Sensing (ASPRS), 210 Little Falls Street, Falls Church, VA 22046-4398, U.S.A.

Tel. 703 534 6617
1. William D. French, Executive Director
Finance: Membership dues, publication sales, convention fees
Purpose: To advance knowledge in photogrammetry and remote sensing
Main Interests: All
Education and Training: Professional certification programme, offering a 'Certified Photogrammetrist' designation to qualified professionals; applicants are peer-judged by a committee. Several meetings are jointly sponsored with other societies, namely the ASPRS Fall Convention and ASPRS/ACSM Annual Meeting. Various other symposia, seminars and workshops are also offered
Publications: *Photogrammetric Engineering and Remote Sensing* (monthly), *Manual of remote sensing*, *SPOT simulation applications handbook*, *Multilingual dictionary of photogrammetry and remote sensing*, *Manual of photogrammetry*, *Extraction of information from remotely sensed images*, *Renewable resources management: applications of remote sensing*, *1985 ASPRS survey of the profession*. Various conference proceedings. Satellite image posters

741 Earth Observation Satellite Company (EOSAT), 4300 Forbes Boulevard, Lanham, MD 20706, U.S.A.
Tel. (301) 552 0500 Telex RCA 277685 LSAT UR
1. C. P. Williams, President
2. R. P. Mroczynski, Director of Public Affairs
Finance: Department of Commerce contract with further revenue from Landsat data sales and fees paid by international receiving stations
Purpose: Operators of Landsat remote sensing satellite system and distributors of all data from these satellites. Primary products include photographic images and tapes from MSS and TM sensors. To provide earth resource data to a worldwide customer base
Main Interests: Landsat earth resources data, its applications and distribution
Education and Training: Informal training offered as seminars by applications specialists. Major presence at conferences sponsored by groups in remote sensing and applications disciplines
Library: Under organization. Landsat data and map base searches. All Landsat MSS, TM and RBV imagery
Publications: *EOSAT Landsat Data Users Notes* (quarterly), *EOSAT Landsat Applications Notes* (bi-monthly), *EOSAT Technical Notes* (irregular)

742 Earth Resources Observation System (EROS) Data Center, National Mapping Division, U.S. Geological Survey, Sioux Falls, SD 57198, U.S.A.
Tel. (605) 594 6511 Telex 910 668 0310
1. D. T. Lauer, Chief, Technique and Application Branch
2. R. Pohl, Chief, Data Production Branch
Finance: Government
Purpose: Responsible for research and development in applications of remotely sensed data and geographic information systems to resource management problems

Main Interests: Applications of aerial photography and Landsat data to resource problems

Education and Training: Training offered annually to Department of Interior staff, discipline-specific groups and on a limited basis the international community. Regularly co-sponsors the Pecora series of conferences

Library: 10,000 volumes, 52 current periodicals; fifteen 16mm films on various aspects of remote sensing are available for loan. EROS is an archive centre for U.S. Geological Survey remotely sensed aircraft and satellite data and holds all Landsat 1, 2, 3, 4 and 5, Shuttle, Skylab, Apollo/Gemini, National High Altitude Photography, NASA and USGS imagery.

743 Technology Application Center, University of New Mexico, 2500 Central SE, Albuquerque, NM 87131, U.S.A.

Tel. 505 277 3622 Telex 660461 ASBKS UNM ABQ

1. Dr. S. A. Morain, Director

Finance: NASA funded

Purpose: Technology transfer in the field of remote sensing of natural resources

Main Interests: Image processing, geographic information systems training, information retrieval and dissemination

Education and Training: Visiting Scientists Program; beginning to advanced-level customized training in remote sensing, usually revolving around a specific topic or theme, content and length dependent on the requirement of the trainee. Short courses, three days' to four weeks' duration, are regularly offered for beginners in remote sensing. Host of the semi-annual meeting of the New Mexico Geographic Information Advisory Committee

Library: TAC has online and in-house access to one of the most complete government document libraries on remote sensing in the United States. Several thousand titles provided by NASA are held on microfiche, some hardcopy on file, but the microfiche documents can be ordered in hardcopy format on request; major remote sensing journals; slide sets and sample air photos for teaching. Over 450 Landsat scenes (U.S., New Mexico and global), 15,000 aerial photographs (New Mexico and Arizona), all hand-held photography (worldwide) from Apollo, Gemini, Skylab, Apollo-Soyuz and Space Shuttle

Publications: *Remote Sensing of Natural Resources: A Quarterly Literature Review*, Newsletter (irregular), publicity material (irregular)

744 Aerial Photography Field Office, 8105 Federal Building, 125 S. State Street, Salt Lake City, Utah, U.S.A.

745 American Congress on Surveying and Mapping (ACSM), 210 Little Falls Street, Falls Church, VA 22046, U.S.A.

746 American Geological Institute, 4220 King Street, Alexandria, VA 22032, U.S.A.

747 American Institute of Aeronautics and Astronautics (AIAA), 555 W. 57th Street, New York, NY 10019, U.S.A.

748 American Society of Civil Engineers, 345 E. 47th Street, New York, NY 10017, U.S.A.

749 American Soil Conservation Service/U.S. Department of Agriculture, 223 W. 2300 Street, Box 30010, Salt Lake City, Utah 84125, U.S.A.

750 Bernice P. Bishop Museum, Library, P.O. Box 19000 A, Honolulu, HI 96817, U.S.A.

751 Center for Earth and Planetary Studies, Regional Planetary Image Facility, National Air and Space Museum, Smithsonian Institution, Washington, DC 20560, U.S.A.

752 Chicago Aerial Survey (CAS) Inc., 2140 Wolf Road, Des Plaines, IL 60018, U.S.A.

753 Continuing Education Administration, 116 Stewart Center, Purdue University, West Lafayette, IN 47907, U.S.A.

754 Daedalus Enterprises Inc., P.O. Box 1869, Ann Arbor, MI 48106, U.S.A.

755 DIPIX Inc., 10220 Old Columbia Road, Rivers Center, Columbia, MD 21046, U.S.A.

756 Earth Satellite Corporation (EARTHSAT), 7222 47th Street, Chevy Chase, MD 20815, U.S.A.

757 Earth Science and Applications Division, 600 Independence Avenue, S.W., Washington, DC 20546, U.S.A.

758 E. Coyote Enterprises Inc., Route 4, Building 228, Box 1119, Mineral Wells, TX 76067, U.S.A.

759 Environmental Research Institute of Michigan (ERIM), P.O. Box 8618, Ann Arbor, MI 48107, U.S.A.

760 GeoRef Document Delivery Service (GDDS), 4220 King Street, Alexandria, VA 22032, U.S.A.

761 Geosat Committee Inc., 153 Kearny Street, Suite 209, San Francisco, CA 94108, U.S.A.

762 Goddard Space Flight Center, Greenbelt, MD 20771, U.S.A.

763 Geoscience and Remote Sensing Society, 345 E. 47th Street, New York, NY 10017, U.S.A.

764 Institute of Electrical and Electronics Engineers Inc., 345 E. 47th Street, New York, NY 10017, U.S.A.

765 Institute for Scientific Information (ISI), 3501 Market Street, Philadelphia, PA 19104, U.S.A.

766 International Mapping Unlimited, 4343 39th Street, N.W., Washington, DC 20016, U.S.A.

767 Jet Propulsion Laboratory (JPL), California Institute of Technology, 4800 Oak Grove Drive, Pasadena, CA 91103, U.S.A.

768 Laboratory for Applications of Remote Sensing (LARS), Purdue University, Entomology Building - Room 214, West Lafayette, IN 47907, U.S.A.

769 Langley Research Center, Mail Stop 103, Hampton, VA 23665, U.S.A.

770 Library of Congress (LOC), Washington, DC 20540, U.S.A.

771 NASA Scientific and Technical Information Branch, Washington, DC 20546, U.S.A.

772 National Aeronautics and Space Administration (NASA), Washington, DC 20546, U.S.A.

773 National Cartographic Information Center (NCIC), 507 National Center, Reston, VA 22092, U.S.A.

774 National Oceanic and Atmospheric Administration (NOAA), National Environmental Satellite Data Information Service (NESDIS), Mundt Federal Building, Sioux Falls, SD 57198, U.S.A.

775 National Space Science Data Center (NSSDC), Code 601, NASA/Goddard Space Flight Center, Greenbelt, MA 20771, U.S.A.

776 Pattern Recognition Society, Georgetown University Medical Centre, 3900 N.W. Reservoir Road, Washington, DC 20007, U.S.A.

777 Pilot Rock Inc., P.O. Box A5, Trinidad, CA 95570, U.S.A.

778 Planetary Image Facility, Brown University, Providence, RI 02912, U.S.A.

779 Planetary Image Facility, Cornell University, Ithaca, NY 14853, U.S.A.

780 Planetary Image Facility, Lunar and Planetary Institute, Houston, TX 77058, U.S.A.

781 Planetary Image Facility, University of Arizona, Tucson, AZ 85721, U.S.A.

782 Planetary Image Facility, University of Hawaii, Honolulu, HI 96882, U.S.A.

783 Planetary Image Facility, U.S. Geological Survey, Flagstaff, AZ 86001, U.S.A.

784 Planetary Image Facility, Washington University, St. Louis, MO 63130, U.S.A.

785 Remote Sensing Enterprises Inc., P.O. Box 2893, La Habra, CA 90681, U.S.A.

786 Remote Sensing Laboratory, University of Kansas Center for Research, 2291 Irving Hill Drive, Campus West, Lawrence, KA 66045, U.S.A.

787 Society for Photo-optical Instrumentation Engineers, P.O. Box 10, Bellingham, WA 98227, U.S.A.

788 Society of Photographic Scientists and Engineers, 7003 Kilworth Lane, Springfield, VA 22151, U.S.A.

789 Spot Image Corporation, 1897 Preston White Drive, Reston, VA 22091, U.S.A.

790 Translations Research Institute, 5914 Pulaski Avenue, Philadelphia, PA 19144, U.S.A.

791 Superintendent of Documents, Government Printing Office, Washington, DC 20402, U.S.A.

792 University Microfilms International (UMI), 300 North Zeeb Road, Ann Arbor, MI 48106, U.S.A.

793 University of Arizona - Microcampus, Box 4, Harvill Building No. 76, Tucson, AZ 85721, U.S.A.

794 University of Georgia Library, Athens, GA 30602, U.S.A.

795 University of Wisconsin, Milwaukee Campus, American Geographical Society, Golda Meir Library, 2377 E. Hartford Avenue, Box 399, Milwaukee, WI 53201, U.S.A.

796 U.S. Geological Survey, Eastern Distribution Branch, 1200 South Eads Street, Arlington, VA 22202, U.S.A.

797 U.S. Geological Survey, National Mapping Division, 104 National Center, 12201 Sunrise Valley Drive, Reston, VA 22092, U.S.A.

798 U.S. Geological Survey, Western Distribution Branch, 169 Federal Center, 1961 Stout Street, Denver, CO 80225, U.S.A.

799 U.S. National Technical Information Service, 5285 Port Royal Road, Springfield, VA 22161, U.S.A.

800 U.S. Patent and Trademark Office, U.S. Department of Commerce, Washington, DC 20231, U.S.A.

801 Virginia State University, Geology Department, Library, Derring Hall, Blacksburg, VA 24061, U.S.A.

802 Washington Remote Sensing Letter, 1057-B National Press Building. Washington, DC 20045, U.S.A.

803 World Bank, International Bank for Reconstruction and Development (IBRD), 1818 High Street, Washington, DC 20433, U.S.A.

804 World Data Center A for Rockets and Satellites, Code 601, NASA/Goddard Space Flight Center, Greenbelt, MA 20771, U.S.A.

Union of Soviet Socialist Republics

805 Akademiia Nauk S.S.S.R., Academy of Sciences of the U.S.S.R., Izdatel'stvo Nauka, Moscow, U.S.S.R.

806 Council on International Co-operation in Research and Uses of Outer Space (INTERCOSMOS), 117901 Moscow, V-71 Leninsky Prospect 14, Moscow, U.S.S.R.

807 Ministerstvo Vysshego i Srendnego Spetsial'nogo Obrazovaniia S.S.S.R.; Izdatel'stvo Instituta Inzhenerov Geodezii, Aerofotos'emki i Kartografii, Moscow 8, U.S.S.R.

808 Moscow Institute of Engineers for Geodesy, Aerial Surveying and Cartography (MIIGAiK), Moscow State University, Moscow, U.S.S.R.

809 National Committee of Photogrammetrists of the U.S.S.R., Chief Administration of Geodesy and Cartography, ul. Krzhizhanovskogo 14, korp 2, Moscow V-218, U.S.S.R.

810 Sojuzkarta, 45 Vovgogradski Pr, Moscow 109 125, U.S.S.R.

811 Vsesoyuznyi Institut Nauchno-Tekhnicheskoi Informatsii (VINITI), Baltiiskaya ul. 14, Moscow A-219, U.S.S.R.

Venezuela

812 Dirección de Cartografía Nacional (National Cartographic Institute), Edf Camejo, Esq. Camejo, Piso 2, Ofc 111, Centro Simón Bolívar, Caracas, Venezuela

813 Fundación Instituto de Ingeniería (CPDI), Edo Mirando Apartado 40200, Caracas 1040 A, Venezuela

Vietnam

814 Committee of Remote Sensing, Vie Khoahoc Vietnam, Nghia Do, Tuliem, Hanoi, Vietnam

Yugoslavia

815 Rudarski Institut Beograd, Remote Sensing Department, Batajnicki put 2, 11081 Zemun, Yugoslavia

Republic of Zaire

816 Institut Géographique du Zaire (Geographical Institute of Zaire), 106 boulevard du 30 juin, B.P. 30 86 Kinshasa, Gombé, Republic of Zaire

817 Programme d'Études des Ressources Terrestres par Satellite (ERTS), B.P. 4834/3092, Kinshasa, Gombé, Republic of Zaire

Zambia

818 Geological Survey Department, P.O. Box 50135, Lusaka, Zambia

Zimbabwe

819 National Committee for Remote Sensing, 8039 Causeway, Harare, Zimbabwe
Tel: 790 701
1. A.B. Made, National Co-ordinator
Finance: Government and external
Purpose: The co-ordination of remote sensing activities in Zimbabwe
Main Interests: The application of remote sensing in resources surveys
Education and Training: Undergraduate and postgraduate courses are held at the University of Zimbabwe
Library: Over 50 Landsat MSS, TM and SPOT scenes for selected areas of Zimbabwe. Some Spacelab photographs are held, and aerial photography of the whole country is flown every five years

Index

All of the publications, products, materials and organizations mentioned in the text or bibliography and directory of Parts II and III are included in this index. Certain abbreviations, acronyms and initialisms are excluded from the index for the sake of brevity; these may be found towards the front of the book in the list of abbreviations and acronyms on pp. ix–xiv.

Throughout the index, numerals in bold-face type refer to entries in the bibliography and directory of organizations, not to page numbers.